Lecture Notes in Artificial Intelligence   8884

Subseries of Lecture Notes in Computer Science

T0212810

Gonzalo A. Aranda-Corral   Jacques Calmet
Francisco J. Martín-Mateos (Eds.)

Artificial Intelligence and Symbolic Computation

12th International Conference, AISC 2014
Seville, Spain, December 11-13, 2014
Proceedings

 Springer

Volume Editors

Gonzalo A. Aranda-Corral
University of Huelva
Department of Information Technologies
Crta. Huelva - Palos de la Frontera, s/n
21810 Palos de la Frontera, Spain
E-mail: aranda@us.es

Jacques Calmet
Karlsruhe Institute of Technology
Institute for Theoretical Informatics
Am Fasanengarten 5
76131 Karlsruhe, Germany
E-mail: calmet@ira.uka.de

Francisco J. Martín-Mateos
University of Seville
Department of Computer Science and Artificial Intelligence
Av. Reina Mercedes, s/n
41012 Seville, Spain
E-mail: fjesus@us.es

ISSN 0302-9743   e-ISSN 1611-3349
ISBN 978-3-319-13769-8   e-ISBN 978-3-319-13770-4
DOI 10.1007/978-3-319-13770-4
Springer Cham Heidelberg New York Dordrecht London

Library of Congress Control Number: 2014955464

LNCS Sublibrary: SL 7 – Artificial Intelligence

Typesetting: Camera-ready by author, data conversion by Scientific Publishing Services, Chennai, India

Printed on acid-free paper

Springer is part of Springer Science+Business Media (www.springer.com)

Preface

This volume is devoted to the proceedings of the 12th International Conference on Artificial Intelligence and Symbolic Computation (AISC 2014). This conference was organized by the University of Seville in Spain and held during December 11–13, 2014, in the Department of Computer Sciences and Artificial Intelligence at the University of Seville.

AISC 2014 was the latest in the series of specialized biennial conferences founded in 1992 by Jacques Calmet and John Campbell. Its goal was to investigate original research contributions in the fields of artificial intelligence (AI) and symbolic computation: Originally, the conferences were called AISMC where the letter M stood for mathematical. From 1998 on the scope became broader than only mathematics and the M was dropped. This trend led to the conference being associated with potentially similar ones, including Calculemus and Mathematical Knowledge Management (MKM), under the integrated federation of the Conference on Intelligent Computer Mathematics (CICM).

From the beginning, the proceedings have appeared as volumes in the LNCS and then the LNAI series.

AISC 2014 was again independent from the CICM federation in order to extend the scope to new domains covered by AI and symbolic computation beyond mathematics. The goals were to bind mathematical domains such as algebraic topology or algebraic geometry to AI, but also to link AI to domains outside pure algorithmic computing.

The new scope thus covers domains that can be described generically as belonging to the mechanization and computability of AI. This covers the basic concepts of computability and new Turing machines, logics including non-classic ones, reasoning, learning, decision support systems but also machine intelligence and epistemology and philosophy of symbolic mathematical computing. Theoretical as well as application papers were solicited.

Among the 22 submitted papers, 16 were selected. Each paper was reviewed by three referees, either members of the Program Committee or sub-reviewers. Acceptance was based solely on the evaluation of the referees and scores achieved. For two accepted papers a revised version was required and refereed again by the same reviewers who evaluated the original submission. In a third case the revised version was rejected.

The scope of the accepted contributions does not fully cover the expected new scope outlined for this series of conferences but illustrates nicely the trend set for these conferences.

October 2014

Gonzalo A. Aranda-Corral
Jacques Calmet
Francisco J. Martín-Mateos

Organization

AISC 2014, the 12th International Conference on Artificial Intelligence and Symbolic Computation, was held at the University of Seville, Seville, December 11-13, 2014. The Department of Computer Science and Artificial Intelligence, University of Seville, was responsible for the organization and local arrangements of the conference.

Conference Direction

Conference chairs

Gonzalo A. Aranda-Corral	University of Huelva
Francisco J. Martín-Mateos	University of Seville

Program Committee

Rajendra Akerkar	Western Norway Research Institute, Norway
Gonzalo A. Aranda-Corral	University of Huelva, Spain
Serge Autexier	DFKI Bremen, Germany
José Antonio Alonso-Jiménez	University of Seville, Spain
Bernhard Beckert	Karlsruhe Institute of Technology, Germany
Belaid Benhamou	University of Aix-Marseille, France
Rudolf Berghammer	Christian-Albrechts-Universität zu Kiel, Germany
Joaquín Borrego-Díaz	University of Seville, Spain
Francisco Botana	University of Vigo, Spain
Jacques Calmet	Karlsruhe Institute of Technology, Germany
Jacques Carette	McMaster University, Canada
John A. Campbell	University College London, UK
Ivo Düntsch	Brock University, Ontario, Canada
Marc Giusti	Ecole Polytechnique, Palaiseau, France
Martin Charles Golumbic	University of Haifa, Israel
Reiner Hähnle	University of Darmstadt, Germany
Tetsuo Ida	University of Tsukuba, Japan
David J. Jeffrey	The University of Western Ontario, Canada
Deepak Kapur	University of New Mexico, USA
Ilias S. Kotsireas	Wilfrid Laurier University, Canada
Luis M. Laita	Royal Academy of Sciences, Madrid, Spain
Francisco J. Martín-Mateos	University of Seville, Spain
Javier Montero	Complutense University of Madrid, Spain

Manuel Ojeda-Aciego	University of Malaga, Spain
Mario Jesus Pérez-Jiménez	University of Seville, Spain
Tomas Recio	University of Cantabria, Spain
Renaud Rioboo	ENSIIE, Evry, France
Eugenio Roanes-Lozano	Complutense University of Madrid, Spain
Julio Rubio	University of la Rioja, Spain
José-Luis Ruiz-Reina	University of Seville, Spain
Fariba Sadri	Imperial College, London, UK
Cristina Sernadas	Technical University of Lisbon, Portugal
Tony Shaska	Oakland University, USA
Volker Sorge	University of Birmingham, UK
Stanley Steinberg	University of New Mexico, USA
Carolyn Talcott	SRI International, USA
Dongming Wang	University Pierre & Marie Curie, Paris, France
Stephen M. Watt	University of Western Ontario, Canada
Michael Wester	University of New Mexico, USA
Wolfgang Windsteiger	RISC, Johannes Kepler University Linz, Austria

Local Arrangements

Pedro Almagro-Blanco	María José Hidalgo-Doblado
José Antonio Alonso-Jiménez	Jaime de Miguel
Joaquín Borrego-Díaz	Antonio Paredes-Moreno
Antonia Chávez-González	M. Carmen Pérez-Cardona
Juan Galán-Páez	José-Luis Ruiz-Reina

Additional Reviewers

Ramón Béjar	Bruce Reznick
James Davenport	Marie Francoise Roy
Antonio E. Flores Montoya	José Javier Segura
Vladimir Klebanov	Mattias Ulbrich
Pavel Pech	Jing Yang
Daniel Plaumann	

Table of Contents

Proving and Computing: Applying Automated Reasoning to the Verification of Symbolic Computation Systems (Invited Talk)

José-Luis Ruiz-Reina

Computational Logic Group
Dept. of Computer Science and Artificial Intelligence, University of Seville
E.T.S.I. Informática, Avda. Reina Mercedes, s/n. 41012 Sevilla, Spain
`jruiz@us.es`

Abstract. The application of automated reasoning to the formal verification of symbolic computation systems is motivated by the need of ensuring the correctness of the results computed by the system, beyond the classical approach of testing. Formal verification of properties of the implemented algorithms require not only to formalize the properties of the algorithm, but also of the underlying (usually rich) mathematical theory.

We show how we can use ACL2, a first-order interactive theorem prover, to reason about properties of algorithms that are typically implemented as part of symbolic computation systems. We emphasize two aspects. First, how we can override the apparent lack of expressiveness we have using a first-order approach (at least compared to higher-order logics). Second, how we can execute the algorithms (efficiently, if possible) in the same setting where we formally reason about their correctness.

Three examples of formal verification of symbolic computation algorithms are presented to illustrate the main issues one has to face in this task: a Gröbner basis algorithm, a first-order unification algorithm based on directed acyclic graphs, and the Eilenberg-Zilber algorithm, one of the central components of a symbolic computation system in algebraic topology.

1   Introduction

Formal verification of the correctness properties of computing systems is one of the main applications of mechanized reasoning. This is applied in any situation where correctness is so important that one has to verify the system beyond the classical approach of testing. For example, safety critical systems or those where failures may produce high economic losses (these include hardware, microprocessors, microcode and software systems or, more precisely, models of them). In these cases, to increase confidence in the system, a theorem stating its main properties is mechanically proved using a theorem prover.

Symbolic computation systems are software systems, so this idea can be applied to formally verify the correctness of the algorithms implemented in them.

G.A. Aranda-Corral et al. (Eds.): AISC 2014, LNAI 8884, pp. 1–6, 2014.

Since in this case they are usually based on a rich mathematical theory, formal verification require not only to formalize the properties of the algorithms, but also of the underlying theory. Another important aspect to take into account is that the models implemented have to be executable and, if possible, efficient.

ACL2 [7,8] is a theorem prover that uses a first-order logic to reason about properties of programs written in an applicative programming language. It has been successfully applied in a number of industrial–strength verification projects [7]. In this talk, we argue that it can be also applied to the verification of symbolic computation systems. We emphasize two aspects. First, how we deal with the apparent lack of expressiveness of a first–order logic (at least compared to higher-order logics). Second, how we can execute the algorithms modeled (efficiently, if possible) in a setting where we also formally reason about them.

We illustrate the main issues one has to deal with when facing this task, by means of three examples. First, Buchberger algorithm for computing a Gröbner basis of a given polynomial ideal. Second, an algorithm, based on directed acyclic graphs, for computing most general unifiers of two given first-order terms. Finally the Eilenberg-Zilber theorem, a central theorem in algebraic topology.

2   The ACL2 System

ACL2 is both a programming language, a logic for reasoning about programs in the language and a theorem prover to assist in the development of proofs of theorems in the logic.

As a programming language, ACL2 is an extension of an applicative subset of Common Lisp. This means that it contains none of Common Lisp that involve side effects like global variables or destructive updates. In this way, functions in the programming language behave as functions in mathematics, and thus one can reason about them using a first-order logic.

The ACL2 logic is a quantifier-free, first-order logic with equality. The logic includes axioms for propositional logic and for a number of Lisp functions and data types, describing the programming language. Rules of inference of the logic include those for propositional calculus, equality and instantiation. But maybe the main rule of inference is the *principle of induction*, that permits proofs by well-founded induction on the ordinal ε_0. This include induction on natural numbers and structural induction.

From the logical point of view, ACL2 functions are total, in the sense that they are defined on every input, even if it is not an intended input. By the *principle of definition*, new function definitions are admitted as definitional axioms only if there exists a measure in which the arguments of each recursive call decrease with respect to a well-founded relation, ensuring in this way that no inconsistencies are introduced by new definitions.

The ACL2 theorem prover mechanizes that logic, being particularly well suited for obtaining mechanized proofs based on simplification and induction. ACL2 is automatic in the sense that once a conjecture is submitted to the prover, the attempt is carried out without interaction with the user. But for non-elementary results, is often the case that the prover fails to find a proof in its first attempt. Thus, we can say that ACL2 is interactive in the sense that the role of the user is essential for a successful use of the prover: usually, she has to provide a number of definition and lemmas (what is called the *logical world*) that the system will use as simplification (rewrite) rules. The proper definitions and lemmas needed for the proof are obtained first from a preconceived hand proof, but also from inspection of failed proof attempts.

3   Gröbner Basis Computation

In [13], a formal verification of a Common Lisp implementation of Buchberger's algorithm [4] for computing Gröbner bases of polynomial ideals is presented. This needed to formalize a number of previous mathematical theories, including results about coefficient fields, polynomial rings and ideals, abstract reductions, polynomial reductions, ordinal measures and of course Gröbner bases. All these notions fit quite well in the first–order ACL2 logic.

It is worth mentioning that this formal project benefited from previous works done in the system. In particular, an ACL2 theory about term rewriting systems had been previously developed [14]. It turns out that the notions of critical pair and of complete rewrite system [2] are closely related to the notions of *s-polynomial* and Gröbner basis, respectively. In fact, some of the results needed in both formalizations are concrete instances of general results about abstract reductions. Thus, once proved the abstract results, they can applied in any concrete context. *Encapsulation* and *functional instantiation* [9] in ACL2 are a good abstraction mechanism that provides some kind of second–order reasoning in this first–order logic, allowing to reuse previous general results in a convenient way.

Another key concept in this formalization is the notion of *polynomial proof*. In this context, a polynomial ideal basis is seen as a rewriting system that reduces polynomial in some sense, generalizing the notion of polynomial division. In [13], the concept of polynomial proof is introduced, as a data structure the contains explicitly all the components of a sequence of polynomial reductions. It turns out that the correctness properties of the Buchberger algorithm can be described as properties of certain functions that transform polynomial proofs.

4   A Dag–Based Quadratic Unification Algorithm

A *unification algorithm* [1] receives as input a pair of first-order terms and returns, whenever it exists, a most general substitution of its variables for terms, such that when applied to both terms, they become equal. Unification is a key component in automated reasoning and in logic programming, for example.

A naive implementation of unification may have exponential complexity in worst cases. Nevertheless, using more sophisticated data structures and algorithms, it is possible to implement unification quadratic in time and linear in space complexity. In [15], the ACL2 implementation and formal verification of such an efficient algorithm is reported. The key idea is to use a data structure based on directed acyclic graphs (*dags*), which allows *structure sharing*. This implementation can be executed in ACL2 at a speed comparable to a similar C implementation, but in addition its correctness properties are formally verified.

Two main issues where encountered in this formalization project:

- For the execution efficiency of the implementation, it is fundamental that the substitution computed by the algorithm is built by iteratively applying destructive updates to the dag representing the terms to unify. In principle, as said before, ACL2 is an applicative programming language, so destructive updates are not allowed. Nevertheless, ACL2 provides *single-threaded objects (stobjs)* [3], which are data structures with a syntactically restricted use, so that only one instance of the object needs ever exist. This means that when executing an algorithm that uses a stobj, its fields can be updated by destructive assignments, while maintaining the applicative semantics for reasoning about it. In this case, using a stobj to store the input terms as a dag (implementing structure sharing by means of pointers), allows to clearly separate reasoning about the logic of the unification process (which is independent of how terms are represented) from the details related to the efficient data structures used.

- In ACL2, functions are total, and before being admitted in the logic, their termination for every possible input has to be proved. Thus, in principle, this unification algorithm cannot be defined in the ACL2 logic, because a possible input could be, for example, a stobj storing a cyclic graph, which could lead the algorithm to a non-terminating execution. Nevertheless, we know that the intended inputs to the algorithm will always be an acyclic graph and that for those intended inputs, it can be proved that the algorithm terminates. Thus, to accept the definition of the algorithm in the logic, we need to introduce in its logical definition a condition checking that the structure stored in the stobj is acyclic and that it represents well-formed terms. Nevertheless, from the efficiency point of view this is unacceptable, since this expensive check would be evaluated in every iteration of the algorithm. Fortunately, the combination of the *defexec* feature [6], together with the *guard verification mechanism* allows to safely skip this expensive check when executing, provided that the function has received a well-formed input and taking into account that it is previously proved that in every iteration the well-formedness condition is preserved.

5   The Eilenberg-Zilber Theorem

The Eilenberg–Zilber theorem [12] is a fundamental theorem in Simplicial Algebraic Topology, establishing a bridge between a geometrical concept (cartesian product) and a algebraic concept (tensor product). Concretely, it states homological equivalence between the cartesian product and the tensor product of two chain complexes. The Eilenberg-Zilber theorem, expressed as a reduction, has a correspondent algorithm that it is a central component of the computer algebra system *Kenzo* [5], devoted to computing in Algebraic Topology.

Since Kenzo is implemented in Common Lisp, it seems that ACL2 is a good choice to verify some of its components, as it is in this case of the Eilenberg-Zilber algorithm. Although Kenzo is far from being implemented using only applicative features, we can formally verify an ACL2 version of the algorithm, and use it as a *verified checker* for the results obtained using Kenzo.

In [11], it is reported a complete ACL2 formal proof of the Eilenberg-Zilber theorem. In fact, the formalization presented was developed reusing part of a previous formal proof of a normalization theorem needed as a preprocessor justifying the way Kenzo works [10].

The formal proof of the Eilenberg-Zilber theorem is not trivial at all. The first issue encountered was that the existing (informal) proofs were not suitable for being formalized in a first–order logic, so a new informal proof had to be carried out by hand, and then formalized in ACL2. In this proof, the key component is a new structure called *simplicial polynomial*, which represent linear combinations of composition of simplicial operators. Although those linear combinations of functions are in principle second–order objects, it turns out that they can be represented as a first-order object. Moreover, the set of simplicial polynomials, together with addition and composition operations, has a ring structure. It turns out that the new proof developed is carried out mainly by establishing a number of lemmas that, although being non-trivial, can be proved by induction and simplification using the ring properties and the properties given by the simplicial identities (the identities that define simplicial sets). Induction and simplification is the kind of reasoning that is suitable for the ACL2 theorem prover.

References

1. Baader, F., Snyder, W.: Unification Theory. In: Handbook of Automated Reasoning. Elsevier Science Publishers (2001)
2. Baader, F., Nipkow, T.: Term Rewriting and All That. Cambridge University Press (1998)
3. Boyer, R., Moore, J S.: Single-Threaded objects in ACL2 (1999), http://www.cs.utexas.edu/users/moore/publications/stobj/main.pdf
4. Buchberger, B.: Bruno Buchberger's PhD thesis 1965: An algorithm for finding the basis elements of the residue class ring of a zero dimensional polynomial ideal. Journal of Symbolic Computation 41(3-4), 475–511 (2006)
5. Dousson, X., Rubio, J., Sergeraert, F., Siret, Y.: The Kenzo Program, Institut Fourier (1999), http://www-fourier.ujf-grenoble.fr/~sergerar/Kenzo/
6. Greve, D., Kaufmann, M., Manolios, P., Moore, J S., Ray, S., Ruiz-Reina, J.-L., Sumners, R., Vroon, D., Wilding, M.: Efficient execution in an automated reasoning environment. Journal of Functional Programming 1, 15–46 (2008)
7. Kaufmann, M., Moore, J S.: ACL2 home page. University of Texas at Austin (2014), http://www.cs.utexas.edu/users/moore/acl2
8. Kaufmann, M., Manolios, P., Moore, J S.: Computer-Aided Reasoning: An Approach. Kluwer Academic Publishers (2000)

9. Kaufmann, M., Moore, J S.: Structured Theory Development for a Mechanized Logic. Journal of Automated Reasoning 26(2), 161–203 (2001)
10. Lambán, L., Martín-Mateos, F.-J., Rubio, J., Ruiz-Reina, J.-L.: Formalization of a normalization theorem in simplicial topology. Annals of Mathematics and Artificial Intelligence 64, 1–37 (2012)
11. Lambán, L., Martín-Mateos, F.-J., Rubio, J., Ruiz-Reina, J.-L.: Verifying the bridge between simplicial topology and algebra: the Eilenberg-Zilber algorithm. Logic Journal of the IGPL 22(1), 39–65 (2014)
12. May, J.P.: Simplicial objects in Algebraic Topology. Van Nostrand (1967)
13. Medina–Bulo, I., Palomo–Lozano, F., Ruiz–Reina, J.-L.: A verified Common Lisp implementation of Buchberger's algorithm in ACL2. Journal of Symbolic Computation 45(1), 96–123 (2010)
14. Ruiz-Reina, J.L., Alonso, J.A., Hidalgo, M.J., Martín-Mateos, F.J.: Formal proofs about rewriting using ACL2. Ann. Math. and Artificial Intelligence 36(3), 239–262 (2002)
15. Ruiz-Reina, J.-L., Alonso, J.-A., Hidalgo, M.-J., Martín-Mateos, F.J.: Formal correctness of a quadratic unification algorithm. Journal of Automated Reasoning 37(1-2), 67–92 (2006)

Combining Systems for Mathematical Creativity (Invited Talk)

Volker Sorge

School of Computer Science
University of Birmingham
V.Sorge@cs.bham.ac.uk
http://www.cs.bham.ac.uk/~vxs

Abstract. We present work on the generation of novel mathematical results by means of integrating heterogeneous reasoning systems.

The combination of reasoning systems has become quite routine in recent years. Most integrations are of a bilateral nature, where one reasoning system draws on the strength of another by incorporating results, delegating subtasks or replaying proofs. This is mostly aimed at supporting standard theorem proving tasks, such as software or hardware verification or the formalisation of existing mathematics.

In this talk we highlight some of our work over the last decade that follows a different approach at system combination: Systems are only loosely integrated and collaboratively perform creative mathematical tasks leading to novel, non-trivial results and theorems. We present a number of techniques developed in the context of problems that are intractable for humans due to large numbers of cases that need to be considered. We describe

- embedding of machine learning into automated theorem proving technology to learn invariants that can uniquely distinguish algebraic entities [5,6],
- the generation of classification theorems in finite algebra both of a quantitative and qualitative nature, i.e., by enumerating equivalence classes and unambiguously describing their elements, using combinations of automated reasoning, SAT solving, machine learning and computer algebra [4,2,7,3],
- and the application of these technologies to support human reasoning for solving problems via infinite case analyses in topology [1].

The work demonstrates not only how the combination of systems can lead to reasoning technology that is more powerful than the sum of its parts, but also that a lightweight approach to combination allows the integration of a plethora of different systems that can efficiently obtain novel mathematical results without foregoing correctness. It also emphasises that the computational component is not a purely ancillary means to achieve a mathematical end, but that there can be a symbiotic relationship, in that the mathematical endeavour leads to advances in automated reasoning technology, while the need to push the scope of automation further can lead to novel mathematical techniques that are interesting results in their own right.

G.A. Aranda-Corral et al. (Eds.): AISC 2014, LNAI 8884, pp. 7–8, 2014.

References

1. Al-Hassani, O., Mahesar, Q.A., Coen, C.S., Sorge, V.: A Term Rewriting System for Kuratowski's Closure-Complement Problem. In: Proceedings of RTA 2012. Leibniz International Proceedings in Informatics, vol. 15. Schloss Dagstuhl (2012)
2. Colton, S., Meier, A., Sorge, V., McCasland, R.: Automatic Generation of Classification Theorems for Finite Algebras. In: Basin, D., Rusinowitch, M. (eds.) IJCAR 2004. LNCS (LNAI), vol. 3097, pp. 400–414. Springer, Heidelberg (2004)
3. Distler, A., Shah, M., Sorge, V.: Enumeration of AG-Groupoids. In: Davenport, J.H., Farmer, W.M., Urban, J., Rabe, F. (eds.) Calculemus/MKM 2011. LNCS, vol. 6824, pp. 1–14. Springer, Heidelberg (2011)
4. Meier, A., Pollet, M., Sorge, V.: Comparing Approaches to the Exploration of the Domain of Residue Classes. Journal of Symbolic Computation 34(4), 287–306 (2002)
5. Meier, A., Sorge, V., Colton, S.: Employing Theory Formation to Guide Proof Planning. In: Calmet, J., Benhamou, B., Caprotti, O., Hénocque, L., Sorge, V. (eds.) AISC-Calculemus 2002. LNCS (LNAI), vol. 2385, pp. 275–289. Springer, Heidelberg (2002)
6. Sorge, V., Meier, A., McCasland, R., Colton, S.: Automatic Construction and Verification of Isotopy Invariants. Journal of Automated Reasoning 40(2-3), 221–243 (2008)
7. Sorge, V., Meier, A., McCasland, R., Colton, S.: Classification Results in Quasigroup and Loop Theory via a Combination of Automated Reasoning Tools. Comm. Mathematicae Universitatis Carolinae 49(2-3), 319–340 (2008)

Models for Logics and Conditional Constraints in Automated Proofs of Termination*

Salvador Lucas[1,2] and José Meseguer[2]

[1] DSIC, Universitat Politècnica de València, Spain
[2] CS Dept. University of Illinois at Urbana-Champaign, IL, USA

Abstract. Reasoning about termination of declarative programs, which are described by means of a *computational logic*, requires the definition of appropriate abstractions as *semantic models* of the logic, and also handling the *conditional constraints* which are often obtained. The formal treatment of such constraints in automated proofs, often using *numeric interpretations* and (arithmetic) constraint solving, can greatly benefit from appropriate techniques to deal with the conditional (in)equations at stake. Existing results from linear algebra or real algebraic geometry are useful to deal with them but have received only scant attention to date. We investigate the definition and use of numeric models for logics and the resolution of linear and algebraic conditional constraints as unifying techniques for proving termination of declarative programs.

Keywords: Conditional constraints, program analysis, termination.

1 Introduction

The operational semantics of sophisticated rule-based programming languages such as CafeOBJ [7], Maude [3], or Haskell [8] is often formalized in a proof-theoretic style by means of a *computational logic*, and the corresponding language interpreters better understood as *inference* machines [15]. The notion of *operational termination* [11] was introduced to give an account of the termination behavior of programs of such languages [12]. An *interpreter* for a logic $\mathcal{L}$ (for instance the logic for *Conditional Term Rewriting Systems* (CTRSs) with inference system in Figure 1) is an *inference machine* that, given a theory $\mathcal{S}$ (e.g., the CTRS $\mathcal{R}$ in Example 1) and a goal formula φ (e.g., a one-step rewriting $s \rightarrow t$ for terms s and t) tries to incrementally build a proof tree for φ by using (instances of) the inference rules $\frac{B_1,\ldots,B_n}{A} \in \mathcal{I}(\mathcal{L})$ of the inference system $\mathcal{I}(\mathcal{L})$ of $\mathcal{L}$. Then, $\mathcal{S}$ is *operationally terminating* if for any φ the interpreter either finds a proof, or fails in all possible attempts (always in finite time). In this setting, practical methods for proving operational termination involve two main issues (see [13] and also [17] for CTRSs): (1) the *simulation* of the (one-step) rewrite relations $\rightarrow$ and $\rightarrow^*$ associated to a CTRS $\mathcal{R}$ and defined by means of the inference system in Figure 1; and (2) the use of (automatically generated) *well-founded relations* $\sqsupset$ to *abstract* rewrite computations and guarantee the absence

---

* Developed during a sabbatical year at UIUC. Supported by projects NSF CNS 13-19109, MINECO TIN2010-21062-C02-02 and TIN 2013-45732-C4-1-P, and GV BEST/2014/026 and PROMETEO/2011/052.

G.A. Aranda-Corral et al. (Eds.): AISC 2014, LNAI 8884, pp. 9–20, 2014.

$$\text{(Refl)} \quad \frac{}{t \to^* t} \qquad \text{(Cong)} \; \frac{s_i \to t_i}{f(s_1, \ldots, s_i, \ldots, s_k) \to f(s_1, \ldots, t_i, \ldots, s_k)}$$
$$\text{for all } k\text{-ary symbols } f \text{ and } 1 \le i \le k$$

$$\text{(Tran)} \quad \frac{s \to u \quad u \to^* t}{s \to^* t} \qquad \text{(Repl)} \; \frac{s_1 \to^* t_1 \; \ldots \; s_n \to^* t_n}{\ell \to r}$$
$$\text{for each rule } \ell \to r \Leftarrow s_1 \to t_1 \cdots s_n \to t_n$$

Fig. 1. Inference rules for the CTRS logic (all variables are universally quantified)

of infinite ones. Here, (1) amounts at dealing with abstractions for sentences like $\forall \boldsymbol{x}(B_1 \wedge \cdots \wedge B_n \Rightarrow A)$ which *simulate* the use of the aforementioned inference rules; and (2) often involves the comparison of expressions s and t using $\sqsupset$, provided that a number of *semantic* conditions (e.g., rewriting steps $s_i \to_{\mathcal{R}}^* t_i$ for some terms s_i and t_i) hold. Abstractions can be formalized as *semantic models* $\mathcal{M} = (\mathsf{D}, \mathcal{F}_\mathsf{D}, \Pi_\mathsf{D})$ of $\mathcal{L}$ (see Section 2), where D is a *domain* and $\mathcal{F}_\mathsf{D}$ and Π_D are interpretations of the function symbols $\mathcal{F}$ and predicates Π of $\mathcal{L}$, respectively. For instance, relations $\to$ and $\to^*$ (which are predicates in the corresponding logic) are typically interpreted as *orderings* on D, often a *numeric* domain like $\mathbb{N}$ or $[0, +\infty)$. In this paper we introduce the idea of using *conditional expressions* to *restrict* such domains in logical models (Section 3). This is often useful.

Example 1. Consider the following CTRS $\mathcal{R}$:

$$\begin{array}{llll}
\mathsf{or}(0, x) \to x & (1) & \mathsf{or}(\mathsf{not}(x), x) \to 1 & (6) & \mathsf{and}(x, \mathsf{not}(x)) \to 0 & (11) \\
\mathsf{or}(x, 0) \to x & (2) & \mathsf{and}(0, x) \to 0 & (7) & \mathsf{and}(\mathsf{not}(x), x) \to 0 & (12) \\
\mathsf{or}(1, x) \to 1 & (3) & \mathsf{and}(x, 0) \to 0 & (8) & \mathsf{not}(1) \to 0 & (13) \\
\mathsf{or}(x, 1) \to 1 & (4) & \mathsf{and}(1, x) \to x & (9) & \mathsf{not}(0) \to 1 & (14) \\
\mathsf{or}(x, \mathsf{not}(x)) \to 1 & (5) & \mathsf{and}(x, 1) \to x & (10) &
\end{array}$$

$$\mathsf{implies}(x, y) \to 1 \Leftarrow \mathsf{not}(x) \to 1 \tag{15}$$
$$\mathsf{implies}(x, y) \to 1 \Leftarrow y \to 1 \tag{16}$$
$$\mathsf{implies}(x, y) \to 0 \Leftarrow x \to 1, y \to 0 \tag{17}$$
$$f(x) \to f(0) \Leftarrow \mathsf{implies}(\mathsf{implies}(x, \mathsf{implies}(x, 0)), 0) \to 1 \tag{18}$$

We *failed* to prove operational termination of $\mathcal{R}$ by using the ordering-based techniques introduced in [13] and also with the more advanced techniques in [14]. However, below we provide a very simple proof of operational termination based on the use (within [13]!) of a *bounded* domain $[0, 1]$ which can be easily implemented by using conditional constraints.

As an interesting specialization of this general idea, Section 4 introduces *convex matrix interpretations* as a new, twofold extension of the framework introduced by Endrullis et al. [5] for TRSs, where rather than using vectors $\boldsymbol{x}$ of natural numbers (or non-negative numbers, as in [2]), we use *convex sets* satisfying a matrix inequality $A\boldsymbol{x} \ge \boldsymbol{b}$. Section 5 discusses existing approaches to deal with the obtained numeric conditional constraints: *Farkas' Lemma* and results from *Algebraic Geometry*. Section 6 compares with related work. Section 7 concludes.

2   Models of Logics for Proofs of Termination

In this paper, $\mathcal{X}$ denotes a set of variables and $\mathcal{F}$ denotes a *signature*: a set of function symbols $\{f, g, \ldots\}$, each with a fixed *arity* given by a mapping $ar : \mathcal{F} \to \mathbb{N}$. The set of *terms* built from $\mathcal{F}$ and $\mathcal{X}$ is $\mathcal{T}(\mathcal{F}, \mathcal{X})$. A CTRS $\mathcal{R} = (\mathcal{F}, R)$ consist of a signature $\mathcal{F}$ and a set R of rules $\ell \to r \Leftarrow s_1 \to t_1, \cdots, s_n \to t_n$, where $l, r, s_1, t_1, \cdots, s_n, t_n \in \mathcal{T}(\mathcal{F}, \mathcal{X})$. For $s, t \in \mathcal{T}(\mathcal{F}, \mathcal{X})$, we write $s \to_\mathcal{R} t$ ($s \to^*_\mathcal{R} t$) if there is a proof for $s \to t$ ($s \to^* t$) with the inference system in Figure 1.

Given a (first order) *logic signature* $\Sigma = (\mathcal{F}, \Pi)$ where $\mathcal{F}$ is a signature of function symbols and Π is a signature of *predicate symbols*, the formulas φ of a (first order) logic $\mathcal{L}$ over Σ are built up from atoms $P(t_1, \ldots, t_k)$ with $P \in \Pi$ and $t_1, \ldots, t_k \in \mathcal{T}(\mathcal{F}, \mathcal{X})$, logic connectives (e.g., $\wedge, \neg, \Rightarrow$) and quantifiers ($\forall, \exists$) in the usual way; $Form_\Sigma$ is the set of such formulas. A *theory* $\mathcal{S}$ of $\mathcal{L}$ is a set of formulas, $\mathcal{S} \subseteq Form_\Sigma$, and its *theorems* are the formulas $\varphi \in Form_\Sigma$ for which we can derive a proof using the inference system $\mathcal{I}(\mathcal{L})$ of $\mathcal{L}$ in the usual way (written $\mathcal{S} \vdash \varphi$). Given a logic $\mathcal{L}$ describing computations in a (declarative) programming language, programs are viewed as *theories* $\mathcal{S}$ of $\mathcal{L}$.

Example 2. In the logic of CTRSs, with binary *predicates* $\to$ and $\to^*$, the theory for a CTRS $\mathcal{R} = (\mathcal{F}, R)$ is obtained from the inference rules in Figure 1 after *specializing* them as $(Cong)_{f,i}$ for each $f \in \mathcal{F}$ and i, $1 \le i \le ar(f)$ and $(Repl)_\rho$ for all $\rho : \ell \to r \Leftarrow c \in R$. Then, inference rules $\frac{B_1, \ldots, B_n}{A}$ become *implications* $B_1 \wedge \cdots \wedge B_n \Rightarrow A$. For instance, for $(Tran)$, $(Cong)_{\mathsf{not}}$, $(Repl)_{(1)}$, and $(Repl)_{(15)}$:

$$\forall s, t, u \ (s \to u \wedge u \to^* t \Rightarrow s \to^* t) \tag{19}$$

$$\forall s, t \ (s \to t \Rightarrow \mathsf{not}(s) \to \mathsf{not}(t)) \tag{20}$$

$$\forall x \ (\mathsf{or}(0, x) \to x) \tag{21}$$

$$\forall x, y \ (\mathsf{not}(x) \to^* 1 \Rightarrow \mathsf{implies}(x, y) \to 1) \tag{22}$$

For analysis and verification purposes we often need to *abstract* $\mathcal{L}$ into a *numeric* setting (e.g., arithmetics, linear algebra, or algebraic geometry) where appropriate techniques are available to *prove* properties of interest. This amounts at giving a (numeric) *model* of $\mathcal{L}$ that *satisfies* $\mathcal{S}$.

An $\mathcal{F}$-algebra is a pair $\mathcal{A} = (\mathsf{D}, \mathcal{F}_\mathsf{D})$, where D is a set and $\mathcal{F}_\mathsf{D}$ is a set of mappings $f_\mathcal{A} : \mathsf{D}^k \to \mathsf{D}$ for each $f \in \mathcal{F}$ where $k = ar(f)$. A Σ-*model* is a triple $\mathcal{M} = (\mathsf{D}, \mathcal{F}_\mathsf{D}, \Pi_\mathsf{D})$ where $(\mathsf{D}, \mathcal{F}_\mathsf{D})$ is an $\mathcal{F}$-algebra, and for each k-ary $P \in \Pi$, $P_\mathcal{M} \in \Pi_\mathsf{D}$ is a k-ary relation $P_\mathcal{M} \subseteq \mathsf{D}^k$. Given a *valuation mapping* $\alpha : \mathcal{X} \to \mathsf{D}$, the evaluation mapping $[\_]^\mathcal{A}_\alpha : \mathcal{T}(\mathcal{F}, \mathcal{X}) \to \mathsf{D}$ (also $[\_]^\mathcal{M}_\alpha$ if $\mathcal{A}$ is part of $\mathcal{M}$) is the unique homomorphism extending α. Finally, $[\_]^\mathcal{M}_\alpha : Form_\Sigma \to Bool$ is given by:

1. $[P(t_1, \ldots, t_k)]^\mathcal{M}_\alpha = \mathsf{true}$ if and only if $([t_1]^\mathcal{M}_\alpha, \ldots, [t_k]^\mathcal{M}_\alpha) \in P_\mathcal{A}$;
2. $[\varphi \wedge \psi]^\mathcal{M}_\alpha = \mathsf{true}$ if and only if $[\varphi]^\mathcal{M}_\alpha = \mathsf{true}$ and $[\psi]^\mathcal{M}_\alpha = \mathsf{true}$;
3. $[\varphi \Rightarrow \psi]^\mathcal{M}_\alpha = \mathsf{true}$ if and only if $[\varphi]^\mathcal{M}_\alpha = \mathsf{false}$ or $[\psi]^\mathcal{M}_\alpha = \mathsf{true}$;
4. $[\neg\varphi]^\mathcal{M}_\alpha = \mathsf{true}$ if and only if $[\varphi]^\mathcal{M}_\alpha = \mathsf{false}$;
5. $[\exists x \, \varphi]^\mathcal{M}_\alpha = \mathsf{true}$ if and only if there is $a \in \mathsf{D}$ such that $[\varphi]^\mathcal{M}_{\alpha[x \mapsto a]} = \mathsf{true}$;
6. $[\forall x \, \varphi]^\mathcal{M}_\alpha = \mathsf{true}$ if and only if for all $a \in \mathsf{D}$, $[\varphi]^\mathcal{M}_{\alpha[x \mapsto a]} = \mathsf{true}$;

We say that $\mathcal{M}$ *satisfies* $\varphi \in Form_\Sigma$ if there is $\alpha \in \mathcal{X} \to D$ such that $[\varphi]_\alpha^{\mathcal{M}} = \text{true}$. If $[\varphi]_\alpha^{\mathcal{M}} = \text{true}$ for *all* valuations α, we write $\mathcal{M} \models \varphi$. A closed formula, i.e., a formula whose variables are all universally or existentially quantified, is called a *sentence*. We say that $\mathcal{M}$ is *a model of a set of sentences* $\mathcal{S} \subseteq Form_\Sigma$ (written $\mathcal{M} \models \mathcal{S}$) if for all $\varphi \in \mathcal{S}$, $\mathcal{M} \models \varphi$. And, given a sentence φ, we write $\mathcal{S} \models \varphi$ if and only if for *all models* $\mathcal{M}$ of $\mathcal{S}$, $\mathcal{M} \models \varphi$. *Sound* logics guarantee that every provable sentence φ is true in *every* model of $\mathcal{S}$, i.e., $\mathcal{S} \vdash \varphi$ implies $\mathcal{S} \models \varphi$.

In practice, $\mathcal{F}$-algebras $\mathcal{A}$ can be obtained if we first consider a *new* set of terms $\mathcal{T}(\mathcal{G}, \mathcal{X})$ where the new symbols $g \in \mathcal{G}$ have 'intended' (often arithmetic) interpretations over an (arithmetic) domain D as mappings g from D into D. The use of the *same name* for the syntactic and semantic objects stresses that they have an *intended* meaning. We associate an expression $e_f \in \mathcal{T}(\mathcal{G}, \{x_1, \ldots, x_k\})$ to each k-ary symbol $f \in \mathcal{F}$, where $x_1, \ldots, x_k \in \mathcal{X}$ are different variables: we write $[f](x_1, \ldots, x_k) = e_f$; and homomorphically extend it to $[\_] : \mathcal{T}(\mathcal{F}, \mathcal{X}) \to \mathcal{T}(\mathcal{G}, \mathcal{X})$. Then, for all $a_1, \ldots, a_k \in D$, we let $f_\mathcal{A}(a_1, \ldots, a_k) = [e_f]_{\alpha_a}$, for α_a given by $\alpha_a(x_i) = a_i$ for all $1 \le i \le k$.

Example 3. For $\mathcal{R}$ in Example 1, $\mathcal{F} = \{0, 1, \text{or}, \text{and}, \text{not}, \text{implies}, f\}$, where $ar(0) = ar(1) = 0$, $ar(f) = 1$, and $ar(\text{or}) = ar(\text{and}) = ar(\text{implies}) = 2$. Let $\mathcal{G} = \{0, 1, max, min, \_ - \_\}$ with $ar(0) = ar(1) = 0$ and $ar(max) = ar(min) = ar(\_ - \_) = 2$. We define an $\mathcal{F}$-algebra over the reals $\mathbb{R}$ as follows:

$$
\begin{array}{llll}
[0] = 0 & [\text{and}](x,y) = min(x,y) & [\text{or}](x,y) = max(x,y) & [f](x) = 0 \\
[1] = 1 & [\text{not}](x) = 1 - x & [\text{implies}](x,y) = max(1 - x, y) &
\end{array}
$$

We define a model $\mathcal{M} = (D, \mathcal{F}_D, \Pi_D)$ if each $P \in \Pi$ is interpreted as a predicate $P_\mathcal{M} \in \Pi_D$, and each $\varphi \in Form_\Sigma$ as a formula $\varphi_\mathcal{M}$, where $\varphi_\mathcal{M} = P_\mathcal{M}([t_1], \ldots, [t_k])$ if $\varphi = P(t_1, \ldots, t_k)$; $\varphi_\mathcal{M} = \varphi_\mathcal{M} \oplus \psi_\mathcal{A}$ if $\varphi = \chi \oplus \psi$ for $\oplus \in \{\wedge, \Rightarrow\}$ and $\varphi_\mathcal{M} = \Box \chi_\mathcal{M}$ if $\varphi = \Box \chi$ for $\Box \in \{\neg, \forall, \exists\}$. The goal is *proving* that $\mathcal{M} \models \mathcal{S}$ holds.

Example 4. We can interpret both $\to$ and $\to^*$ as '=' (intended to be the equality among real numbers). Then, the sentences in Example 2 become

$$\forall s, t, u \in \mathbb{R} \ (s = u \wedge u = t \Rightarrow s = t) \tag{23}$$

$$\forall s, t \in \mathbb{R} \ (s = t \Rightarrow 1 - s = 1 - t) \tag{24}$$

$$\forall x \in \mathbb{R} \ (max(0, x) = x) \tag{25}$$

$$\forall x, y \in \mathbb{R} \ (1 - x = 1 \Rightarrow max(1 - x, y) = 1) \tag{26}$$

Unfortunately, (25) and (26) do *not* hold in the intended model due to the (big) algebraic domain $\mathbb{R}$. For instance, $max(0, -1) = 0 \neq -1$, i.e., (25) is *not* true.

Example 4 shows that the appropriate definition of the *domain* of a model is crucial to satisfy a set of formulas. The next section investigates this problem.

3   Domains for Algebras and Models Revisited

In proofs of termination, domains D for numeric $\mathcal{F}$-algebras $\mathcal{A}$ usually are infinite (subsets of) n-dimensional open intervals which are bounded from below: $\mathbb{N}^n$ or

$[0, +\infty)^n$ for some $n \geq 1$. Furthermore, considered orderings often make the corresponding ordered sets *total* (like $[0, +\infty)$ ordered by $\geq_{\mathbb{R}}$), or nontotal but with subsets $B \subseteq D$ bounded by some value $x_B \in D$ (like $[0, +\infty)^n$ ordered by the pointwise extension of the usual ordering $\geq_{\mathbb{R}}$ over the reals, which is a complete lattice). More general domains can be often useful, though.

Example 5. (Continuing Example 4) Although (23) and (24) always hold (under the *intended* interpretation of '=' as the equality), *satisfiability* of other sentences may depend on the considered *domain* of values: if $D = [0, 1]$, then (25) and (26) hold; if $D = \mathbb{N}$, then only (25) holds. The use of $D = [0, 1]$ can be made *explicit* in (25) and (26) by adding further constraints:

$$\forall x \in \mathbb{R}\ (\ x \geq 0 \wedge 1 \geq x \Rightarrow max(0, x) = x)\ \ (27)$$
$$\forall x, y \in \mathbb{R}\ (\ x \geq 0 \wedge 1 \geq x \wedge y \geq 0 \wedge 1 \geq y \wedge 1 - x = 1 \Rightarrow max(1 - x, y) = 1)\ \ (28)$$

Thus, we need to deal with *conditional constraints* for using such more general domains. Also to handle *max* expressions [6,16].

Example 6. We can *expand* the definition of *max* in (27) and (28) into:

$$\forall x \in \mathbb{R}\ (\ x \geq 0 \wedge 1 \geq x \wedge 0 \geq x \Rightarrow 0 = x)\ \ (29)$$
$$\forall x \in \mathbb{R}\ (\ x \geq 0 \wedge 1 \geq x \wedge x \geq 0 \Rightarrow x = x)\ \ (30)$$
$$\forall x, y \in \mathbb{R}\ (\ x \geq 0 \wedge 1 \geq x \wedge y \geq 0 \wedge 1 \geq y \wedge 1 - x = 1 \wedge 1 - x \geq y \Rightarrow 1 - x = 1)\ \ (31)$$
$$\forall x, y \in \mathbb{R}\ (\ x \geq 0 \wedge 1 \geq x \wedge y \geq 0 \wedge 1 \geq y \wedge 1 - x = 1 \wedge y > 1 - x \Rightarrow y = 1)\ \ (32)$$

where (30) clearly holds true and we do not longer care about it.

3.1 Conditional Domains for Term Algebras and Models

Given a set D and a predicate χ over D, we let $D_\chi = \{x \in D \mid \chi(x)\}$ be the *restriction* of D by χ. An $\mathcal{F}$-algebra $\mathcal{A} = (D, \mathcal{F}_D)$ yields a *restricted* $\mathcal{F}$-algebra $\mathcal{A}_\chi = (D_\chi, \mathcal{F}_{D_\chi})$, where for each $f \in \mathcal{F}$, $f_{\mathcal{A}_\chi}$ is the restriction of $f_\mathcal{A}$ to D_χ^k, if for all k-ary symbols $f \in \mathcal{F}$, this *algebraicity* or *closedness* condition holds:

$$\forall x_1, \ldots, x_k\ \left(\left(\bigwedge_{i \leq i \leq k} \chi(x_i) \right) \Rightarrow \chi(f_\mathcal{A}(x_1, \ldots, x_k)) \right) \quad (33)$$

guaranteeing that if $f_\mathcal{A}$ is given inputs in D_χ, the outcome belongs to D_χ as well.

Remark 1. Algebraicity is a standard requirement for algebraic interpretations. Most times, however, the imposition of simple requirements on the shape of the numeric expressions e_f used to define $f_\mathcal{A}$ (see Section 2) makes this task easy and often avoids any checking. A well-known example is taking $\mathcal{D} = [0, +\infty)$ and requiring e_f to be a *polynomial* whose coefficients are all *non-negative*.

The relations $P_\mathcal{M} \subseteq D^k$ interpreting k-ary predicates $P \in \Pi$ can be restricted to $P_{\mathcal{M}_\chi} = P_\mathcal{M} \cap D_\chi^k$ to yield a new interpretation of P in $\mathcal{M}_\chi = (D_\chi, \mathcal{F}_{D_\chi}, \Pi_{D_\chi})$. For practical purposes, in this paper we only consider simple restrictions of $\mathcal{F}$-algebras and models, where D is obtained as the solution of *linear* constraints.

Definition 1 (Convex Polytopic Domain). *Given a matrix* $A \in \mathbb{R}^{m \times n}$, *and* $\boldsymbol{b} \in \mathbb{R}^m$, *the set of solutions of the inequality* $A\boldsymbol{x} \geq \boldsymbol{b}$ *is a* convex polytope $D(A, \boldsymbol{b}) = \{\boldsymbol{x} \in \mathbb{R}^n \mid A\boldsymbol{x} \geq \boldsymbol{b}\}$. *We call* $D(A, \boldsymbol{b})$ *a convex polytopic domain.*

Example 7. For $A = (-1, 1)^T$ and $\boldsymbol{b} = (-1, 0)$, we have $D(A, \boldsymbol{b}) = [0, 1]$. If $A = (1)$ and $\boldsymbol{b} = (0)$, then $D(A, \boldsymbol{b}) = [0, +\infty)$.

Example 8. Continuing Example 3, we obtain an $\mathcal{F}$-algebra $\mathcal{A}_{[0,1]} = ([0, 1], \mathcal{F}_{[0,1]})$ as the restriction to $[0, 1]$ of the $\mathcal{F}$-algebra over $\mathbb{R}$ defined there. The constraints (25) and (26) are written in the restricted model as follows

$$\forall x \in [0, 1] \, (max(0, x) = x) \tag{34}$$

$$\forall x, y \in [0, 1] \, (1 - x = 1 \Rightarrow max(1 - x, y) = 1) \tag{35}$$

After encoding memberships like $x \in [0, 1]$ as inequalities $x \geq 0 \land 1 \geq x$ and expanding the definition of max, we obtain $(29) - (32)$.

In sharp contrast with Example 4, restricting the model at hand to $[0, 1]$ leads to a model for $\mathcal{R}$ in Example 1 which is useful to prove its operational termination.

Example 9. According to [13], $\mathcal{R}$ in Example 1 is operationally terminating if there is a relation $\gtrsim$ on terms such that $\to^* \subseteq \gtrsim$, and a well-founded ordering $\sqsupset$ satisfying $\gtrsim \circ \sqsupset \subseteq \sqsupset$ such that, for all substitutions σ, if $\sigma(\mathsf{implies}(\mathsf{implies}(x, \mathsf{implies}(x, 0)), 0)) \to_{\mathcal{R}}^* \sigma(1)$ holds, then $\sigma(\mathsf{F}(x)) \sqsupset \sigma(\mathsf{F}(0))$ for the rule (*dependency pair*[1]) $\mathsf{F}(x) \to \mathsf{F}(0) \Leftarrow \mathsf{implies}(\mathsf{implies}(x, \mathsf{implies}(x, 0)), 0) \to 1$ (where F is a fresh symbol). Let $\mathcal{M} = ([0, 1], \mathcal{F}'_{[0,1]}, \Pi_{[0,1]})$ where $\mathcal{F}' = \mathcal{F} \cup \{F\}$, $\mathcal{F}'_{[0,1]}$ is $\mathcal{F}_{[0,1]}$ as in Example 8 extended with $[\mathsf{F}](x) = x$, and $\Pi_{[0,1]}$ given by $\to_{[0,1]} = \to_{[0,1]}^* = (=_{[0,1]})$ (i.e., the equality on $[0, 1]$). $\mathcal{M}$ is a model of $\mathcal{R}$; by soundness, if $s \to^* t$ holds for $s, t \in \mathcal{T}(\mathcal{F}, \mathcal{X})$, we have $[s] =_{[0,1]} [t]$. Let $\gtrsim$ be as follows: for all $s, t \in \mathcal{T}(\mathcal{F}, \mathcal{X})$, $s \gtrsim t$ holds if and only if $[s] =_{[0,1]} [t]$. Then, $\to^* \subseteq \gtrsim$, as desired.

Now, consider the ordering $>_1$ over $\mathbb{R}$ given by $x >_1 y$ if and only if $x - y \geq 1$; it is a well-founded relation on $[0, 1]$ (see [10]). We let $\sqsupset$ be the (well-founded) relation on $\mathcal{T}(\mathcal{F}, \mathcal{X})$ induced by $>_1$ as before. Again, for all substitutions σ, if $\sigma(\mathsf{implies}(\mathsf{implies}(x, \mathsf{implies}(x, 0)), 0)) \to_{\mathcal{R}}^* \sigma(1)$ holds, then, by soundness,

$$[\sigma(\mathsf{implies}(\mathsf{implies}(x, \mathsf{implies}(x, 0)), 0))] =_{[0,1]} [\sigma(1)] \tag{36}$$

holds as well. We also have

$$\forall x \in [0, 1]([\mathsf{implies}(\mathsf{implies}(x, \mathsf{implies}(x, 0)), 0)] =_{[0,1]} [1] \Rightarrow [\mathsf{F}(x)] >_1 [\mathsf{F}(0)]) \tag{37}$$

because, for all $x \geq 0$,

$$
\begin{aligned}
[\mathsf{implies}(\mathsf{implies}(x, \mathsf{implies}(x, 0)), 0)] &= max(1 - max(1 - x, max(1 - x, 0)), 0) \\
&= max(1 - max(1 - x, 1 - x), 0) \\
&= max(1 - (1 - x), 0) \\
&= max(x, 0) \\
&= x
\end{aligned}
$$

---

[1] For the purpose of this paper, the procedure to obtain this new rule is not relevant. The interested reader can find the details in [13].

and hence $[\mathsf{implies}(\mathsf{implies}(x, \mathsf{implies}(x, 0)), 0)] =_{[0,1]} [1]$ holds only if $x = 1 = [1]$. Combining (36) and (37), we conclude that, for all substitutions σ, if $\sigma(\mathsf{implies}(\mathsf{implies}(x, \mathsf{implies}(x, 0)), 0)) \rightarrow^*_{\mathcal{R}} \sigma(1)$ holds, then $[\mathsf{F}(x)]^{\mathcal{M}} = 1 >_1 0 = [\mathsf{F}(0)]^{\mathcal{M}}$, as desired. This proves operational termination of $\mathcal{R}$ in Example 1.

In the following section we discuss an interesting application of convex polytopic domains to improve the well-known matrix interpretations [5,2].

4 Convex Matrix Interpretations

A *convex matrix intepretation* for a k-ary symbol f is a linear expression $F_1 x_1 + \cdots + F_k x_k + F_0$, where $F_1, \ldots, F_k \in \mathbb{R}^{n \times n}$ are (square) matrices, $F_0 \in \mathbb{R}^n$ and $x_1, \ldots, x_k \in \mathbb{R}^n$, which is closed on $D(A, b)$, i.e., that satisfies

$$\forall x_1, \ldots x_k \in \mathbb{R}^n \left(\bigwedge_{i=1}^{k} A x_i \geq b \Rightarrow A(F_1 x_1 + \cdots + F_k x_k + F_0) \geq b \right) \quad (38)$$

An $\mathcal{F}$-algebra $\mathcal{A} = (\mathsf{D}, \mathcal{F}_\mathsf{D})$ is obtained if $\mathsf{D} = D(A, b)$, and each k-ary symbol $f \in \mathcal{F}$ is given $f_\mathcal{A}(x_1, \ldots, x_k) = F_1 x_1 + \cdots + F_k x_k + F_0$ that satisfies (38). The following ordering $\geq$ is considered: $x = (x_1, \ldots, x_n) \geq (y_1, \ldots, y_n) = y$ if $x_i \geq y_i$ for all $1 \leq i \leq n$. Given $\delta > 0$, the (strict) ordering $>_\delta$ is also used: $x = (x_1, \ldots, x_n) >_\delta (y_1, \ldots, y_n) = y$ if $x_1 - y_1 \geq \delta$ and $(x_2, \ldots, x_n) \geq (y_2, \ldots, y_n)$.

Remark 2. Convex matrix interpretations include the usual matrix interpretations in [5,2] if $A = I_{n \times n}$ and $b = \mathbf{0} \in \mathbb{R}^n$.

In contrast to $(\mathbb{N}, \geq)$ and $([0, +\infty), \geq)$, that are *total* orders, and also to $(\mathbb{N}^n, \geq)$ and $([0, +\infty)^n, \geq)$, that are not total, but are complete lattices, $(D(A, b), \geq)$ does *not* need to be total or a complete lattice. This has some interesting advantages.

Example 10. Consider the CTRS $\mathcal{R}$ [17, Example 7.2.45]:

$$a \rightarrow a \Leftarrow b \rightarrow x, c \rightarrow x \quad (39) \qquad\qquad c \rightarrow d \Leftarrow d \rightarrow x, e \rightarrow x \quad (41)$$
$$b \rightarrow d \Leftarrow d \rightarrow x, e \rightarrow x \quad (40)$$

According to [13], $\mathcal{R}$ is operationally terminating if there is a relation $\gtrsim$ such that $\rightarrow^* \subseteq \gtrsim$, and $\sqsupset$ is a well-founded ordering such that $\gtrsim \circ \sqsupset \subseteq \sqsupset$ and for the *dependency pair* $a^\sharp \rightarrow a^\sharp \Leftarrow b \rightarrow x, c \rightarrow x$ (for $a^\sharp$ a new symbol), we have that, for all substitutions σ, if $b \rightarrow^* \sigma(x)$ and $c \rightarrow^* \sigma(x)$, then $a^\sharp \sqsupset a^\sharp$. With

$$A = \begin{bmatrix} 1 & 1 \\ 1 & 0 \\ 0 & 1 \end{bmatrix} \text{ and } b = (1, 0, 0)^T, \text{ together with:}$$

$$[a] = [a^\sharp] = \begin{bmatrix} 1 \\ 0 \end{bmatrix} \qquad [b] = [d] = \begin{bmatrix} 1 \\ 0 \end{bmatrix} \qquad [c] = [e] = \begin{bmatrix} 0 \\ 1 \end{bmatrix}$$

we have $[a], [a^\sharp], [b], [c], [d], [e] \in D(A, b)$, as required by (38). It can be proved that $(D(A, b), \mathcal{F}_{D(A,b)}, \Pi_{D(A,b)})$, where $\rightarrow, \rightarrow^* \in \Pi$ are both interpreted (in

$\Pi_{D(A,b)}$) as $\geq$ is a *model* of $\mathcal{R}$. For $s, t \in \mathcal{T}(\mathcal{F}, \mathcal{X})$, we let $s \gtrsim t$ if and only if $[s] \geq [t]$. Thus, $\rightarrow^* \subseteq \gtrsim$ holds. The ordering $>_1$ on $D(A, b)$ is *well-founded* because $[0, +\infty)$ is bounded from below (see [10]). Thus, for $s, t \in \mathcal{T}(\mathcal{F}, \mathcal{X})$, we define $s \sqsupset t$ if and only if $[s] >_1 [t]$. Now, since $\rightarrow^* \subseteq \gtrsim$, we only have to prove that $[b] \geq [x] \wedge [c] \geq [x] \Rightarrow [a^\sharp] >_1 [a^\sharp]$, i.e.,

$$\forall x_1, x_2 \in \mathbb{R} \left(\begin{bmatrix} 1 & 1 \\ 1 & 0 \\ 0 & 1 \end{bmatrix} \begin{bmatrix} x_1 \\ x_2 \end{bmatrix} \geq \begin{bmatrix} 1 \\ 0 \\ 0 \end{bmatrix} \wedge \begin{bmatrix} 1 \\ 0 \end{bmatrix} \geq \begin{bmatrix} x_1 \\ x_2 \end{bmatrix} \wedge \begin{bmatrix} 0 \\ 1 \end{bmatrix} \geq \begin{bmatrix} x_1 \\ x_2 \end{bmatrix} \Rightarrow \begin{bmatrix} 1 \\ 0 \end{bmatrix} >_1 \begin{bmatrix} 1 \\ 0 \end{bmatrix} \right) \quad (42)$$

which can be written as a universally quantified conjunction of two formulas:

$$x_1 + x_2 \geq 1 \wedge x_1 \geq 0 \wedge x_2 \geq 0 \wedge 1 \geq x_1 \wedge 0 \geq x_2 \wedge 0 \geq x_1 \wedge 1 \geq x_2 \Rightarrow 1 >_1 1 \quad (43)$$
$$x_1 + x_2 \geq 1 \wedge x_1 \geq 0 \wedge x_2 \geq 0 \wedge 1 \geq x_1 \wedge 0 \geq x_2 \wedge 0 \geq x_1 \wedge 1 \geq x_2 \Rightarrow 0 \geq 0 \quad (44)$$

The crucial point is that the conditional part of the implications does not hold because no $x \in D(A, b)$ satisfies $(1, 0)^T \geq x$ and $(0, 1)^T \geq x$ (see Example 11).

The following sections discuss existing mathematical techniques that can be used to automatically deal with the conditional constraints obtained so far.

5   Conditional Polynomial Constraints

In this section, we explore well-known results from linear algebra [20] and algebraic geometry [18] to deal with conditional polynomial constraints.

5.1   Conditional Constraints with Linear Polynomials

Farkas' Lemma provides a *(universal) quantifier elimination* result for linear (conditional) sentences (cf. [20]).

Theorem 1 (Affine form of Farkas' Lemma). *Let $Ax \geq b$ be a linear system of k inequalities and n unknowns over the real numbers with non-empty solution set S and let $c \in \mathbb{R}^n$ and $\beta \in \mathbb{R}$. Then, the following statements are equivalent:*

1. $c^T x \geq \beta$ *for all* $x \in S$,
2. $\exists \lambda \in \mathbb{R}_0^k$ *such that* $c = A^T \lambda$ *and* $\lambda^T b \geq \beta$.

By condition (1) in Theorem 1 proving $\forall x \, (Ax \geq b \Rightarrow c^T x \geq \beta)$ can be recast as the *constraint solving problem* of finding a nonnegative vector λ such that c is a linear nonnegative combination of the *rows* of A and β is smaller than the corresponding linear combination of the components of b. Note that if $Ax \geq b$ has no solution, i.e., S in Theorem 1 is *empty*, the conditional sentence trivially holds. Thus, we do not need to check S for emptiness when using Farkas' result.

Example 11. Sentences (43) and (44) can be proved using Theorem 1. This proves operational termination of $\mathcal{R}$ in Example 10.

Example 12. After encoding the equality as the conjunction of $\geq$ and $\leq$, we transform sentences (29), (31) and (32) into:

$$\forall x \in \mathbb{R} \, (\, x \geq 0 \wedge 1 \geq x \wedge 0 \geq x \Rightarrow 0 \geq x) \tag{45}$$

$$\forall x \in \mathbb{R} \, (\, x \geq 0 \wedge 1 \geq x \wedge 0 \geq x \Rightarrow x \geq 0) \tag{46}$$

$$\forall x, y \in \mathbb{R} \, (\, x \geq 0 \wedge 1 \geq x \wedge y \geq 0 \wedge 1 \geq y \wedge 1 - x = 1 \wedge 1 - x \geq y \Rightarrow 1 - x \geq 1) \tag{47}$$

$$\forall x, y \in \mathbb{R} \, (\, x \geq 0 \wedge 1 \geq x \wedge y \geq 0 \wedge 1 \geq y \wedge 1 - x = 1 \wedge 1 - x \geq y \Rightarrow 1 \geq 1 - x) \tag{48}$$

$$\forall x, y \in \mathbb{R} \, (\, x \geq 0 \wedge 1 \geq x \wedge y \geq 0 \wedge 1 \geq y \wedge 1 - x = 1 \wedge y > 1 - x \Rightarrow y \geq 1) \tag{49}$$

$$\forall x, y \in \mathbb{R} \, (\, x \geq 0 \wedge 1 \geq x \wedge y \geq 0 \wedge 1 \geq y \wedge 1 - x = 1 \wedge y > 1 - x \Rightarrow 1 \geq y) \tag{50}$$

which are conditional linear sentences provable using Farkas' Lemma.

5.2 Conditional Constraints with Arbitrary Polynomials

Given polynomials $h_1, \ldots, h_m \in \mathbb{R}[X_1, \ldots, X_n]$, the *semialgebraic set* defined by $h_1, \ldots, h_m$ is $W_{\mathbb{R}}(h) = W_{\mathbb{R}}(h_1, \ldots, h_m) = \{x \in \mathbb{R}^n \mid h_1(x) \geq 0 \wedge \cdots \wedge h_m(x) \geq 0\}$. A well-known *representation theorem* establishes that a polynomial which is positive for all tuples $(x_1, \ldots, x_n) \in W_{\mathbb{R}}(h)$ can be written as a linear combination of $h_1, \ldots, h_m$ with 'coefficients' s that are sums of squares of polynomials ($s \in \sum \mathbb{R}[X]^2$) [18, Theorem 5.3.8]. If we can write a polynomial f as a linear combination of $h_1, \ldots, h_m$ with 'coefficients' that are sums of squares, this provides a *certificate* of non-negativeness of f on $W_{\mathbb{R}}(h_1, \ldots, h_m)$: sums of squares are non-negative, all h_i are non-negative on values in $W_{\mathbb{R}}(h_1, \ldots, h_m)$ and the product and addition of non-negative numbers is non-negative. Explicitly:

Theorem 2. *Let* $\mathbb{R}[X] := \mathbb{R}[X_1, \ldots, X_n]$, $h_1, \ldots, h_m \in \mathbb{R}[X]$, $W_{\mathbb{R}}(h) = W_{\mathbb{R}}(h_1, \ldots, h_m)$ *and* $S \subseteq \mathbb{R}$ *such that* $W_{\mathbb{R}}(h_1, \ldots, h_m) \subseteq S^n$. *Let* $s_i \in \sum \mathbb{R}[X]^2$ *for all* i, $0 \leq i \leq m$. *If for all* $x_1, \ldots, x_n \in S$, $f \geq s_0 + \sum_{i=1}^{m} s_i \cdot h_i$, *then, for all* $(x_1, \ldots, x_n) \in W_{\mathbb{R}}(h_1, \ldots, h_m)$, $f(x_1, \ldots, x_n) \geq 0$.

Example 13. Consider the constraint $X_1 \geq X_2^2 \wedge X_2 \geq X_3^2 \Rightarrow X_1 \geq X_3^4$ from [16, page 51]. With $s_0 = (X_3^2 - X_2)^2$, $s_1 = 1$ and $s_2 = 2X_3^2$, we have:

$$X_1 - X_3^4 = (X_3^2 - X_2)^2 + (X_1 - X_2^2) + 2X_3^2 \cdot (X_2 - X_3^2)$$

witnessing that the constraint holds.

6 Related Work

The material in Section 2 can be thought of as a generalization and extension of the *intepretation method* for proving termination of Term Rewriting Systems (see, e.g., [17, Section 5]). The interpretation method uses *ordered algebras* which are algebras $\mathcal{A}$ with domain D including one or more ordering relations $\succeq_{\mathsf{D}}, \succ_{\mathsf{D}}$, etc., satisfying a number of properties (stability, monotonicity, etc.). Such relations are used to *induce* relations $\succeq, \succ$ on terms which are then used to compare the left- and right-hand sides ℓ and r of rewrite rules $\ell \to r$. The targeted rules in such comparisons and the conclusions we may reach depend on the considered approach for proving termination (see [17, Sections 5.2 and 5.4], for instance). In our setting, orderings are introduced as interpretations of computational relations (e.g., → and →*), and we do not require anything special about them beyond their ability to provide a *model* of the theory at hand. For instance, where the interpretation method requires *monotonicity*, we just expect the relation to provide a model of rules (*Cong*), which encode the monotonicity of the rewrite relation. The advantage is that we do not need reformulations of the framework when other logics are considered; in contrast, the interpretation method requires explicit adaptations. For instance, in *Context-Sensitive Rewriting* [9] rewritings are propagated to selected arguments of function symbols only. Thus, (*Cong*) may have *no specialization* for *some* arguments i of some symbols f. Whereas this requires specific adaptations of the interpretation method (see, e.g., [21]), we can apply our methods without any change. Furthermore, although our practical examples involve CTRSs, our development does not really depend on that and applies to arbitrary declarative languages.

With regard to existing approaches to deal with conditional constraints in proofs of termination, the following result formalizes the transformational approach to deal with polynomial *conditional constraints* in [6,16].

Proposition 1. [16, Proposition 3] *Let prem and conc be two polynomials with natural coefficients, where conc is not a constant. Let $p_1, \ldots, p_{m+1}, q_1, \ldots, q_{m+1}$ be arbitrary polynomials with natural coefficients. If*

$$conc(p_{m+1}) - conc(q_{m+1}) - prem(p_1, \ldots, p_m) + prem(q_1, \ldots, q_m) \geq 0$$

is valid over the natural numbers, then $p_1 \geq q_1 \wedge \cdots \wedge p_m \geq q_m \Rightarrow p_{m+1} \geq q_{m+1}$ is also valid over the natural numbers.

This result holds if *prem* and *conc* have non-negative real coefficients, and variables range over nonnegative real numbers. When linear polynomials are used this technique is subsumed by Farkas' lemma.

Proposition 2. *Let $C \in \mathbb{R}_{\geq 0}[Y]$ and $P \in \mathbb{R}_{\geq 0}[Y_1, \ldots, Y_m]$ be linear applications with C nonconstant, i.e., $C = \gamma Y$ with $\gamma > 0$ and $P = \sum_{i=1}^{m} \pi_i Y_i$. Let $p_i, q_i \in \mathbb{R}_{\geq 0}[X_1, \ldots, X_n]$ be linear polynomials for all i, $1 \leq i \leq m+1$, i.e., $p_i = p_{i0} + \sum_{j=1}^{n} p_{ij}X_j$ and $q_i = q_{i0} + \sum_{j=1}^{n} q_{ij}X_j$. Let $A = (p_{ij} - q_{ij})_{m,n}$, $\mathbf{b} = (q_{10} - p_{10}, \ldots q_{m0} - p_{m0})^T$, $\mathbf{c} = (p_{m+1,1} - q_{m+1,1}, \ldots, p_{m+1,n} - q_{m+1,n})^T$ and $\beta = q_{m+1,0} - p_{m+1,0}$. If for all $X_1, \ldots, X_m \geq 0$, $C(p_{m+1}) - C(q_{m+1}) - P(p_1, \ldots, p_m) + P(q_1, \ldots, q_m) \geq 0$, then there is $\boldsymbol{\lambda} \in \mathbb{R}_0^m$ such that $\mathbf{c} \geq A^T \boldsymbol{\lambda}$ and $\beta \leq \boldsymbol{\lambda}^T \mathbf{b}$.*

Remark 3. Regarding mechanization, Nguyen et al.'s technique has a drawback with respect to those in Section 5. Given a rule $\ell \to r \Leftarrow \bigwedge_{i=1}^{n} s_i \to t_i$, Nguyen et al.'s technique requires that both $[s_i]$ and $[t_i]$ are polynomials *with non-negative coefficients only*. This is because $[s_i]$ and $[t_i]$ are handled separately by polynomials *conc* and *prem*. But in an implementation, $[s_i]$ and $[t_i]$ are *parametric polynomials* where the coefficients are *parameters* rather than numbers (see [4,10] for instance). Thus, we need to *constrain* them to be *non-negative* in order to use the technique. In contrast, we do not restrict the coefficients of polynomials in any way. Hence, the coefficients of the parametric polynomials could be negative numbers without any problem. For instance, this is crucial to synthesize $D(A, \boldsymbol{b}) = [0, 1]$ used in the examples above, where A and $\boldsymbol{b}$ require negative numbers.

Farkas' Lemma is used in proofs of termination of *imperative* programs in [19].

7 Conclusion

We have provided a generic, logic-oriented approach to abstraction in proofs of termination of programs in declarative languages, which is based on defining appropriate *models* for logics. We have used numeric domains defined as *restrictions* of 'big' numeric sets by means of predicates that can be handled as conditional constraints. We have introduced *convex* domains and used them to extend the powerful *matrix interpretation method* for proving termination of TRSs in two directions: the use of *convex* domains and the application to other logics (e.g., CTRSs). We have shown the usefulness of these general purpose ideas by applying them to prove operational termination of CTRSs: $\mathcal{R}$ in Example 1 could *not* be handled within the recently introduced 2D DP framework for proving operational termination of CTRSs [13] or its extensions [14]; but the weakness was not in the framework itself, but in the available algebraic interpretations: we can prove $\mathcal{R}$ operationally terminating now due to the use of a convex domain like $[0, 1]$. And powerful tools like AProVE do not find a proof of operational termination of $\mathcal{R}$ in Example 10 by using transformations. In contrast, we found a simple proof with convex matrix interpretations and the techniques in [13].

We have shown that existing, powerful techniques to deal with numeric constraints provide an appropriate framework for implementing the previous techniques. We have implemented most of these techniques as part of our tool MU-TERM [1]. In particular, the use of Farkas' Lemma for dealing with linear conditional constraints obtained from linear polynomial interpretations and matrix interpretations plays a central role in the implementation of the 2D DP framework for operational termination of CTRSs [13] which is presented in [14]. In [10, Example 13], we advocate the use of *negative* coefficients in proofs of termination of *CSR* using polynomial interpretations. The implementation, though, was tricky (see [10, Sections 6.1.3 and 7]). This paper is a step forward because: (1) our treatment is valid for arbitrary polynomials. We do not need to provide special results as [10, Observation 1] to deal with polynomials of some specific form (quadratic, cubic, ...); (2) we avoid the introduction of disjunctive constraints which lead to an exponential blowup and to an expensive constraint solving process; and (3) we admit negative numbers everywhere. They are treated as any other number and there is no need to 'assert' which of the coefficients could be negative in order to handle them apart (see [10, Section 7] and [10, Example 20]). However, much work is necessary to make fully general use of these techniques in practical applications. We plan to address these issues in the near future.

References

1. Alarcón, B., Gutiérrez, R., Lucas, S., Navarro-Marset, R.: Proving Termination Properties with MU-TERM. In: Johnson, M., Pavlovic, D. (eds.) AMAST 2010. LNCS, vol. 6486, pp. 201–208. Springer, Heidelberg (2011)
2. Alarcón, B., Lucas, S., Navarro-Marset, R.: Using Matrix Interpretations over the Reals in Proofs of Termination. In: Proc. of PROLE 2009, pp. 255–264 (2009)
3. Clavel, M., Durán, F., Eker, S., Lincoln, P., Martí-Oliet, N., Meseguer, J., Talcott, C. (eds.): All About Maude - A High-Performance Logical Framework. LNCS, vol. 4350. Springer, Heidelberg (2007)
4. Contejean, E., Marché, C., Tomás, A.-P., Urbain, X.: Mechanically proving termination using polynomial interpretations. J. of Aut. Reas. 34(4), 325–363 (2006)
5. Endrullis, J., Waldmann, J., Zantema, H.: Matrix Interpretations for Proving Termination of Term Rewriting. J. of Aut. Reas. 40(2-3), 195–220 (2008)
6. Fuhs, C., Giesl, J., Middeldorp, A., Schneider-Kamp, P., Thiemann, R., Zankl, H.: Maximal Termination. In: Voronkov, A. (ed.) RTA 2008. LNCS, vol. 5117, pp. 110–125. Springer, Heidelberg (2008)
7. Futatsugi, K., Diaconescu, R.: CafeOBJ Report. AMAST Series. World Scientific (1998)
8. Hudak, P., Peyton-Jones, S.J., Wadler, P.: Report on the Functional Programming Language Haskell: a non–strict, purely functional language. Sigplan Notices 27(5), 1–164 (1992)
9. Lucas, S.: Context-sensitive computations in functional and functional logic programs. Journal of Functional and Logic Programming 1998(1), 1–61 (1998)
10. Lucas, S.: Polynomials over the reals in proofs of termination: from theory to practice. RAIRO Theoretical Informatics and Applications 39(3), 547–586 (2005)
11. Lucas, S., Marché, C., Meseguer, J.: Operational termination of conditional term rewriting systems. Information Processing Letters 95, 446–453 (2005)
12. Lucas, S., Meseguer, J.: Proving Operational Termination of Declarative Programs in General Logics. In: Proc. of PPDP 2014, pp. 111–122. ACM Digital Library (2014)
13. Lucas, S., Meseguer, J.: 2D Dependency Pairs for Proving Operational Termination of CTRSs. In: Proc. of WRLA 2014. LNCS, vol. 8663 (to appear, 2014)
14. Lucas, S., Meseguer, J., Gutiérrez, R.: Extending the 2D DP Framework for CTRSs. In: Selected papers of LOPSTR 2014. LNCS (to appear, 2015)
15. Meseguer, J.: General Logics. In: Ebbinghaus, H.-D., et al. (eds.) Logic Colloquium 1987, pp. 275–329. North-Holland (1989)
16. Nguyen, M.T., de Schreye, D., Giesl, J., Schneider-Kamp, P.: Polytool: Polynomial interpretations as a basis for termination of logic programs. Theory and Practice of Logic Programming 11(1), 33–63 (2011)
17. Ohlebusch, E.: Advanced Topics in Term Rewriting. Springer (April 2002)
18. Prestel, A., Delzell, C.N.: Positive Polynomials. In: From Hilbert's 17th Problem to Real Algebra. Springer, Berlin (2001)
19. Podelski, A., Rybalchenko, A.: A Complete Method for the Synthesis of Linear Ranking Functions. In: Steffen, B., Levi, G. (eds.) VMCAI 2004. LNCS, vol. 2937, pp. 239–251. Springer, Heidelberg (2004)
20. Schrijver, A.: Theory of linear and integer programming. John Wiley & Sons (1986)
21. Zantema, H.: Termination of Context-Sensitive Rewriting. In: Comon, H. (ed.) RTA 1997. LNCS, vol. 1232, pp. 172–186. Springer, Heidelberg (1997)

Using Representation Theorems
for Proving Polynomials Non-negative*

Salvador Lucas

DSIC, Universidad Politécnica de Valencia
Camino de Vera s/n, 46022 Valencia, Spain

Abstract. Proving polynomials non-negative when variables range on a subset of numbers (e.g., $[0, +\infty)$) is often required in many applications (e.g., in the analysis of *program termination*). Several *representations* for *univariate* polynomials P that are non-negative on $[0, +\infty)$ have been investigated. They can often be used to *characterize* the property, thus providing a method for checking it by trying a *match* of P against the representation. We introduce a new characterization based on viewing polynomials P as *vectors*, and find the appropriate *polynomial basis* $\mathcal{B}$ in which the non-negativeness of the *coordinates* $[P]_{\mathcal{B}}$ representing P in $\mathcal{B}$ witnesses that P is non-negative on $[0, +\infty)$. Matching a polynomial against a representation provides a way to transform universal sentences $\forall x \in [0, +\infty)\ P(x) \geq 0$ into a *constraint solving* problem which can be solved by using efficient methods. We consider different approaches to solve both kind of problems and provide a quantitative evaluation of performance that points to an early result by Pólya and Szegö's as an appropriate basis for implementations in most cases.

Keywords: Polynomial constraints, positive polynomials, representation theorems.

1 Introduction

Representations of univariate polynomials that are positive $(Pd(I))$ or non-negative $(Psd(I))$ on an *interval* I of real numbers have been investigated (see [14] for a survey) and some of them are useful to *check* the property. In this paper we investigate this question: *which technique is worth to be implemented for a practical use?* Our specific motivation is the development of *efficient* and *automatic* tools for proving *termination* of programs, where polynomials play a prominent role (see [8,12], for instance) and the focus is on $Psd([0, +\infty))$.

We decompose the whole problem into two main steps: (1) the use of representation theorems to obtain a set of *existential constraints* whose satisfaction witnesses that $(\forall x \geq 0)\ P \geq 0$ holds and (2) the use of constraint solving techniques to obtain appropriate solutions. With regard to (1), several researchers (starting with Hilbert) addressed this problem and contributed in different ways (see Section 2). In this setting, the following test is often used in practice [10]: a

---

* Developed during a sabbatical year at UIUC. Supported by projects NSF CNS 13-19109, MINECO TIN2010-21062-C02-02 and TIN 2013-45732-C4-1-P, and GV BEST/2014/026 and PROMETEO/2011/052.

G.A. Aranda-Corral et al. (Eds.): AISC 2014, LNAI 8884, pp. 21–33, 2014.

polynomial P is $Psd([0, +\infty)^n)$ if *all coefficients of the monomials in P are non-negative*. This has obvious limitations. For instance, $Q(x) = x^3 - 4x^2 + 6x + 1$ is $Psd([0, +\infty))$, but contains negative coefficients. The following observation generalizes this approach (Section 3): $P \in \mathbb{R}[X]$ of degree n can be *represented* as a *vector* $[P]_\mathcal{B} = (\alpha_0, \ldots, \alpha_n)^T$ of $n+1$ *coordinates* with respect to a *basis* $\mathcal{B} = \{v_0, \ldots, v_n\} \subseteq \mathbb{R}[X]$, i.e., $P = \alpha_0 v_0 + \alpha_1 v_1 + \cdots + \alpha_n v_n$. Then, P is $Psd([0, +\infty))$ if (i) $[P]_\mathcal{B} \geq \mathbf{0}$ and (ii) $v_0, \ldots, v_n$ are $Psd([0, +\infty))$. Requiring all coefficients in the representation $P = \sum_{i=0}^n p_i x^i$ to be non-negative corresponds to considering the *standard* basis $\mathcal{S}_n = \{1, x, \ldots, x^n\}$ for polynomials of degree n. In our running example, $[Q]_{\mathcal{S}_3} = (1, 6, -4, 1)^T \not\geq \mathbf{0}$. We define a *parametric* polynomial basis $\mathcal{P}_n$ such that, for all $P \in \mathbb{R}[X]$ of degree n which is $Psd([0, +\infty))$, $[P]_\mathcal{B} \geq \mathbf{0}$ for some specific $\mathcal{B}$ which is obtained from $\mathcal{P}_n$ by giving appropriate values to the parameters. We also show how to give value to the parameters.

Example 1. The representation of $Q(x) = x^3 - 4x^2 + 6x + 1$ with respect to $\mathcal{B} = \{1, x, x^2, x(x-2)^2\}$ is $[Q]_\mathcal{B} = (1, 2, 0, 1)^T \geq \mathbf{0}$.

Regarding (2), in Section 4 we use a recent, efficient procedure to solve polynomial constraints over finite domains [5] as a reference to provide a quantitative analysis of the characterizations discussed in Sections 2 and 3 and provide an answer to our question. Section 5 discusses some related work and concludes.

2 Representation of Polynomials Non-negative in $[0, +\infty)$

We consider the following representations of $Psd([0, +\infty))$ polynomials P (see [14]): (1) Hilbert [9]; (2) Pólya and Szegö [13]; (3) Karlin and Studden [11]; and (4) Hilbert's approach using Gram matrices [7].

Remark 1. Our motivation for considering these particular methods is that, in automatic proofs of termination, polynomials P whose non-negativity must be guaranteed are *parametric*, i.e., the coefficients are *not* numbers but rather variables whose value is *generated* by a constraint solving process. All previous methods fit the requirement of being amenable to this practical setting.

We briefly discuss how to use these four methods and also give some cost indicators: $V(n)$ is the number of *parameters* used to match P (of degree n) against the representation, and $I(n)$ is the number of *(in)equalities* which are obtained. The following fact is used later.

Proposition 1. *Let $P, Q \in \mathbb{R}[X_1, \ldots, X_n]$ be $P = \sum_\alpha a_\alpha X^\alpha$ and $Q = \sum_\alpha b_\alpha X^\alpha$. If $a_\alpha \geq b_\alpha$ for all $\alpha \in \mathbb{N}^n$ and Q is $Psd([0, +\infty)^n)$, then P is $Psd([0, +\infty)^n)$.*

In the following, $\div$ and $\%$ denote the integer division and remainder, respectively. We say that a polynomial P is a *sum of squares* (or just *sos*, often denoted as $P \in \sum \mathbb{R}[\mathbf{X}]^2$) if can be written $P = \sum_i f_i^2$ for polynomials f_i.

2.1   Hilbert

Since $P \in \mathbb{R}[X_1, \ldots, X_n]$ is $Psd([0, +\infty)^n)$ if and only if $H(X_1, \ldots, X_n) = P(X_1^2, \ldots, X_n^2)$ is $Psd(\mathbb{R}^n)$ (note that this transformation *doubles* the degree of P), we can use the following result.

Proposition 2 (Hilbert). [9] *If $P \in \mathbb{R}[X]$ is $Psd(\mathbb{R})$, then P is a sum of two squares of polynomials.*

Example 2. Consider $H(x) = Q(x^2) = x^6 - 4x^4 + 6x^2 + 1 = f_1(x) + f_2(x)$ where $f_i(x) = (a_i x^3 + b_i x^2 + c_i x + d_i)^2$ for $i = 1, 2$. Then, $H(x)$ should match

$$\sum_{i=1}^{2} a_i^2 x^6 + 2a_i b_i x^5 + (b_i^2 + 2a_i c_i)x^4 + 2(b_i c_i + a_i d_i)x^3 + (2b_i d_i + c_i^2)x^2 + 2c_i d_i x + d_i^2$$

which amounts at *solving* the following *equalities*:

$$\sum_{i=1}^{2} a_i^2 = 1 \qquad \sum_{i=1}^{2} a_i b_i = 0 \qquad \sum_{i=1}^{2} b_i^2 + 2a_i c_i = -4$$
$$\sum_{i=1}^{2} b_i c_i + a_i d_i = 0 \qquad \sum_{i=1}^{2} 2b_i d_i + c_i^2 = 6 \qquad \sum_{i=1}^{2} c_i d_i = 0 \qquad \sum_{i=1}^{2} d_i^2 = 1$$

A solution (with *irrational* numbers) is obtained by using, e.g., *Mathematica*. We have $V(n) = 2n + 2$ and $I(n) = 2n + 1$.

2.2   Pólya and Szegö

Proposition 3 (Pólya & Szegö). [13] *If P is $Psd([0, +\infty))$, then there are sos polynomials f, g such that $P(x) = f(x) + xg(x)$ and $\deg(f), \deg(xg) \leq \deg(P)$.*

If $f, g \in \sum \mathbb{R}[X]^2$, then both f and xg are $Psd([0, +\infty))$. Thus, Pólya and Szegö's representation actually provides a *characterization*. We can use it, then, to *prove* that P is $Psd([0, +\infty))$ iff P *matches* the representation. Since every univariate *sos* polynomial f can be written as a sum of *two* squares of polynomials, in Proposition 3 we assume $f = f_1^2 + f_2^2$ and $g = g_1^2 + g_2^2$, for polynomials f_i and g_i, $i = 1, 2$. If $n = \deg(P) = 1$, then, since $\deg(f), \deg(xg) \leq 1$, $f, g \in \sum \mathbb{R}[X]^2$ must be *constant* polynomials $f = f_0$ and $g = g_0$. If $n = 2$, then, since $\deg(xg) \leq 2$, $g \in \sum \mathbb{R}[X]^2$ must be a *constant*. If $n > 2$, then $\deg(f_i) = d_1 \leq \lfloor \frac{n}{2} \rfloor$, and $\deg(g_i) = d_2 \leq \lfloor \frac{n-1}{2} \rfloor$. Write $f_i = a_{i,d_1} x^{d_1} + \cdots + a_{i,1} x + a_{i,0}$ and $g_i = b_{i,d_2} x^{d_2} + \cdots + b_{i,1} x + b_{i,0}$ for $i = 1, 2$. Try to *match* the coefficients of the *target* polynomial P against this representation.

Example 3. For our running example Q, we have

$$Q(x) = x^3 - 4x^2 + 6x + 1 = f_1(x) + f_2(x) + x(g_1(x) + g_2(x))$$

where $f_i(x) = (a_i x + b_i)^2$ and $g_i(x) = (c_i x + d_i)^2$ for $i = 1, 2$. Then,

$$Q(x) = (c_1^2 + c_2^2)x^3 + (a_1^2 + a_2^2 + 2c_1 d_1 + 2c_2 d_2)x^2 + (2a_1 b_1 + 2a_2 b_2 + d_1^2 + d_2^2)x + b_1^2 + b_2^2$$

By Proposition 1, rather than equalities, we solve now the *inequalities*[1]:

$$1 \geq c_1^2 + c_2^2; \; -4 \geq a_1^2 + a_2^2 + 2c_1 d_1 + 2c_2 d_2; \; 6 \geq 2a_1 b_1 + 2a_2 b_2 + d_1^2 + d_2^2; \; 1 \geq b_1^2 + b_2^2.$$

with: $a_1 = 0$, $a_2 = 0$, $b_1 = 1$, $b_2 = 0$, $c_1 = 1$, $c_2 = 0$, $d_1 = -2$, and $d_2 = 1$.

---

[1] Using inequalities makes the constraint solving process more flexible and often avoids the use of *irrational* numbers, often out of the scope for most constraint solving tools.

Each f_i and g_i contributes with $d_1 + 1$ and $d_2 + 1$ parametric coefficients, respectively, i.e., $V(n) = 2(d_1 + 1 + d_2 + 1) = 2(2 + d_1 + d_2) = 2(n + 1) = 2n + 2$. The number of *inequalities* to be solved is $I(n) = n + 1$ (one per coefficient p_i of P).

2.3   Karlin and Studden

Theorem 1 (Karlin and Studden). [11, Corollary V.8.1] *Let P_{2m} be a polynomial of degree $2m$ for some $m \geq 0$ with leading coefficient $a_{2m} > 0$. If P_{2m} is $Pd([0, +\infty))$, then there exists a unique representation*

$$P_{2m}(X) = a_{2m} \prod_{j=1}^{m} (X - \alpha_j)^2 + \beta X \prod_{j=2}^{m} (X - \gamma_j)^2$$

where $\beta > 0$ and $0 = \gamma_1 < \alpha_1 < \gamma_2 < \cdots < \gamma_m < \alpha_m < \infty$. Similarly, if P_{2m+1} is a polynomial of degree $2m + 1$ for some $m \geq 0$, with leading coefficient $a_{2m+1} > 0$ and P_{2m+1} is $Pd([0, +\infty))$, then there exists a unique representation

$$P_{2m+1}(X) = a_{2m+1} X \prod_{j=2}^{m+1} (X - \alpha_j)^2 + \beta \prod_{j=1}^{m} (X - \gamma_j)^2$$

where $\beta > 0$ and $0 = \alpha_1 < \gamma_1 < \alpha_2 < \gamma_2 < \cdots < \gamma_m < \alpha_{m+1} < \infty$.

Unfortunately, this representation *cannot* be used to *prove* that P is $Pd([0, +\infty))$ by *matching*. For instance, $P = (x - 1)^2$ matches it, but it is *not* $Pd([0, +\infty))$. However, Karlin and Studden's representation can be used to prove P to be $Psd([0, +\infty))$ by *matching* if we just require $\alpha_j, \beta, \gamma_j \geq 0$.

Example 4. Since the degree of Q is odd, we let

$$K_Q(x) = x(x - \alpha_2)^2 + \beta(x - \gamma_1)^2 = x^3 + (\beta - 2\alpha_2)x^2 + (\alpha_2^2 - 2\beta\gamma_1)x + \beta\gamma_1^2$$

Thus, we have the following constraints (using Proposition 1):

$$-4 \geq \beta - 2\alpha_2 \quad 1 \geq 0 \quad 6 \geq \alpha_2^2 - 2\beta\gamma_1 \quad 1 \geq \beta\gamma_1^2 \quad \beta \geq 0 \quad \gamma_1 \geq 0 \quad \alpha_2 \geq 0$$

The assignment $\alpha_2 = \frac{9}{4}$, $\beta = \frac{1}{4}$, and $\gamma_1 = \frac{1}{2}$ solves the system.

We have $V(n) = n$ and $I(n) = n + 1 + V(n) = 2n + 1$.

2.4   Hilbert with Gram Matrices

An alternative way to use Hilbert's representation is the following.

Theorem 2. [7] *Let P be a polynomial of degree $2m$ and $z(X)$ be the vector of all monomials X^α such that $|\alpha| \leq m$. Then, P is a sum of squares in $\mathbb{R}[X]$ if and only if there exists a real, symmetric, psd matrix B such that $P = z(X)^T B z(X)$.*

Proving $H(x) = P(x^2)$ of degree $2n$ to be *sos* amounts at (1) matching H against $z(X)^T B z(X)$ (where $z(X) = (1, X, \ldots, X^n)^T$) and (2) proving $B \in \mathbb{R}^{n+1 \times n+1}$ positive semidefinite. Since B is symmetric, we need $\frac{(n+1)(n+2)}{2}$ parameters b_{ij} to represent B. Then, we need to solve $2n + 1$ equations in $\frac{(n+1)(n+2)}{2}$ variables (the parameters b_{ij}) corresponding to the monomials in H. According to [15], this can be done by taking $\frac{(n+1)(n+2)}{2} - (2n+1) = \frac{n^2-n}{2}$ of the b_{ij} as *unknowns* which can be given appropriate values that are obtained using (2), i.e., B must be positive semidefinite. This can be done by computing the characteristic polynomial $det(zI_{n+1} - B) = \sum_{i=0}^{n} c_i z^i$ of B and requiring its roots to be non-negative [15]. They show that this can be achieved by imposing $(-1)^{i+n+1} c_i \geq 0$ for all $0 \leq i \leq n$. Thus, $V(n) = \frac{(n+1)(n+2)}{2}$. and $I(n) = (2n+1) + (n+1) = 3n+2$.

3   Checking Positiveness of Polynomials as Vectors

Let V be an n-dimensional vector space over the reals and $\mathcal{B} = \{v_1, \ldots, v_n\}$ be an ordered basis for V. For all n-tuples $\boldsymbol{\alpha} = (\alpha_1, \ldots, \alpha_n) \in \mathbb{R}^n$ we write $\boldsymbol{\alpha} \geq 0$ if $\alpha_i \geq 0$ and $\boldsymbol{\alpha} > 0$ if $\alpha_1 > 0$ and $\alpha_2, \ldots, \alpha_n \geq 0$. Every $v \in V$ can be represented as a *coordinate* vector $[v]_{\mathcal{B}} = (\alpha_1, \ldots, \alpha_n)^T \in \mathbb{R}^n$ such that $v = \alpha_1 v_1 + \cdots + \alpha_n v_n$. Given bases $\mathcal{B}$ and $\mathcal{B}'$ for V, there is a *change of base* matrix (cb-matrix) $M_{\mathcal{B}' \mapsto \mathcal{B}}$ (or just M) which can be used to obtain the *coordinate* representation $[v]_{\mathcal{B}}$ of v in $\mathcal{B}$ from the representation $[v]_{\mathcal{B}'}$ of v in $\mathcal{B}'$: $[v]_{\mathcal{B}} = M[v]_{\mathcal{B}'}$. The set $\mathbb{P}_n$ of *univariate* polynomials of degree at most n is a vectorial space of dimension $n+1$ and has a *standard basis* $\mathcal{S}_n = \{1, x, \ldots, x^n\}$. If $\mathcal{B} = \{v_0, \ldots, v_n\}$ is a basis for $\mathbb{P}_n$ and every $v \in \mathcal{B}$ is $Psd([0, +\infty))$, then given $P \in \mathbb{P}_n$, if $[P]_{\mathcal{B}} = (\alpha_0, \ldots, \alpha_n)^T \geq \mathbf{0}$, then P is $Psd([0, +\infty))$. If $P = \sum_{i=0}^{n} p_i x^i$, this is translated into the *search* of a basis $\mathcal{B}$ satisfying the conditions above and a cb-matrix $M = M_{\mathcal{S}_n \mapsto \mathcal{B}}$ such that $M[P]_{\mathcal{S}_n} \geq \mathbf{0}$. We consider *parametric* bases $\mathcal{B}$ consisting of polynomials with *parametric* coefficients which can be given appropriate values as to fit the requirements above. By a *parametric* polynomial we mean a polynomial $P \in \mathbb{R}[\gamma_1, \ldots, \gamma_k][X]$ over X whose monomials have *coefficients* in $\mathbb{R}[\gamma_1, \ldots, \gamma_k]$; variables $\gamma_1, \ldots, \gamma_k$ are called *parameters*. For all $i \in \mathbb{N}$, consider the parametric univariate polynomials, :

$$\mathsf{P}_i(x) = \prod_{j=1}^{\frac{i}{2}} (x - \gamma_{ij})^2 \text{ if } i \text{ is even} \qquad \mathsf{P}_i(x) = x \prod_{j=1}^{\frac{i-1}{2}} (x - \gamma_{ij})^2 \text{ if } i \text{ is odd}$$

where the empty product is 1, and γ_{ij} are *parameters* satisfying $\gamma_{ij} \geq 0$. For instance, $\mathsf{P}_0(x) = 1$, $\mathsf{P}_1(x) = x$, $\mathsf{P}_2(x) = (x - \gamma_{21})^2 = \gamma_{21}^2 - 2\gamma_{21} x + x^2$, and $\mathsf{P}_3(x) = x(x - \gamma_{31})^2 = \gamma_{31}^2 x - 2\gamma_{31} x^2 + x^3$. Note that for all $i \geq 0$ and $x \geq 0$, $\mathsf{P}_i(x) \geq 0$ and $\mathsf{P}_0(x) > 0$. Given $n \in \mathbb{N}$, let $\mathcal{P}_n = \{\mathsf{P}_0(x), \ldots, \mathsf{P}_n(x)\}$ *ordered* by the sequence $0, 1, \ldots, n$. $\mathcal{P}_n$ is a basis of $\mathbb{P}_n$; this is a consequence of the following.

Theorem 3. *Let* $\mathcal{P} = \{P_0, \ldots, P_n\}$ *be a set of* $n+1$ *polynomials such that* $P_0 \in \mathbb{R} - \{0\}$ *and* $deg(P_i) = i$ *for all* $1 \leq i \leq n$. *Then,* $\mathcal{P}$ *is a basis of* $\mathbb{P}_n(x)$.

Note that $\mathcal{P}_{n+1} = \mathcal{P}_n \cup \mathsf{P}_{n+1}(x)$.

Proposition 4 (Number of Parameters in the Basis). *Given* $n \in \mathbb{N}$, *the number* $N(n)$ *of parameters in* $\mathcal{P}_n$ *is given by* $N(0) = 0$ *and* $N(n) = N(n-1) + \lfloor \frac{n}{2} \rfloor$ *for* $n > 0$. *Furthermore,* $N(n) = \frac{n^2}{4}$ *if* n *is even and* $\frac{n^2-1}{4}$ *otherwise.*

We prove that $\mathcal{P}_n$ *characterizes* $Psd([0, +\infty))$ and $Pd([0, +\infty))$.

Theorem 4. *A polynomial* $P \in \mathbb{R}[X]$ *of degree* n *is* $Psd([0, +\infty))$ *($Pd([0, +\infty))$)* *if and only if* $[P]_{\mathcal{P}_n} \geq \mathbf{0}$ *(resp.* $[P]_{\mathcal{P}_n} > \mathbf{0}$*) for some assignment of values* $\gamma_{ij} \geq 0$ *to the parameters in* $\mathcal{P}_n$.

We show how to compute the cb-matrix $M_n = M_{\mathcal{S}_n \mapsto \mathcal{P}_n}$ for obtaining the representation $[P]_{\mathcal{P}_n} = M_n[P]_{\mathcal{S}_n}$ of $P \in \mathbb{P}_n$ which is required in Theorem 4. In the following, $[\mathsf{P}_n(x)]_{\mathcal{S}_n}^{1,\dots,n}$ is the n-dimensional vector containing the first n (parametric) coordinates of $[\mathsf{P}_n(x)]_{\mathcal{S}_n}$ (the last one is 1, corresponding to x^n).

Theorem 5 (Incremental cb-matrix). *We have* $M_0 = I_1$ *and for all* $n > 0$,

$$M_n = \begin{pmatrix} M_{n-1} & -M_{n-1}[\mathsf{P}_n(x)]_{\mathcal{S}_n}^{1,\dots,n} \\ \mathbf{0}_{1 \times n} & 1 \end{pmatrix}$$

Example 5. Since $M_1 = I_2$, according to Theorem 5, we have:

$$M_2 = \begin{pmatrix} M_1 & -M_1 \begin{pmatrix} \gamma_{21}^2 \\ -2\gamma_{21} \end{pmatrix} \\ \mathbf{0}_{1 \times 2} & 1 \end{pmatrix} = \begin{pmatrix} 1 & 0 & -\gamma_{21}^2 \\ 0 & 1 & 2\gamma_{21} \\ 0 & 0 & 1 \end{pmatrix} \quad \text{and}$$

$$M_3 = \begin{pmatrix} M_2 & -M_2 \begin{pmatrix} 0 \\ \gamma_{31}^2 \\ -2\gamma_{31} \end{pmatrix} \\ \mathbf{0}_{1 \times 3} & 1 \end{pmatrix} = \begin{pmatrix} 1 & 0 & -\gamma_{21}^2 & -2\gamma_{21}^2\gamma_{31} \\ 0 & 1 & 2\gamma_{21} & 4\gamma_{21}\gamma_{31} - \gamma_{31}^2 \\ 0 & 0 & 1 & 2\gamma_{31} \\ 0 & 0 & 0 & 1 \end{pmatrix}$$

For our running example $[Q]_{\mathcal{S}_3} = (1, 6, -4, 1)^T$, we *impose* $[Q]_{\mathcal{P}_3} = M_3[Q]_{\mathcal{S}_3} > 0$:

$$\begin{pmatrix} 1 & 0 & -\gamma_{21}^2 & -2\gamma_{21}^2\gamma_{31} \\ 0 & 1 & 2\gamma_{21} & 4\gamma_{21}\gamma_{31} - \gamma_{31}^2 \\ 0 & 0 & 1 & 2\gamma_{31} \\ 0 & 0 & 0 & 1 \end{pmatrix} \begin{pmatrix} 1 \\ 6 \\ -4 \\ 1 \end{pmatrix} = \begin{pmatrix} 1 + 4\gamma_{21}^2 - 2\gamma_{21}^2\gamma_{31} \\ 6 - 8\gamma_{21} + 4\gamma_{21}\gamma_{31} - \gamma_{31}^2 \\ -4 + 2\gamma_{31} \\ 1 \end{pmatrix} > \begin{pmatrix} 0 \\ 0 \\ 0 \\ 0 \end{pmatrix}$$

The corresponding existential constraint:
$$\gamma_{21}, \gamma_{31} \geq 0, \ 1 + 4\gamma_{21}^2 - 2\gamma_{21}^2\gamma_{31} > 0 \wedge 6 - 8\gamma_{21} + 4\gamma_{21}\gamma_{31} - \gamma_{31}^2 \geq 0 \wedge 2\gamma_{31} - 4 \geq 0 \wedge 1 > 0$$
is satisfied if $\gamma_{21} = 0$ and $\gamma_{31} = 2$, witnessing Q as $pd([0, +\infty))$ through the coordinate representation $[Q]_{\mathcal{P}_3} = (1, 2, 0, 1)^T$ when $\mathcal{P}_3 = \{1, x, x^2, x(x - 2)^2\}$.

Note that $V(n) = N(n) = \frac{n^2 - n\%2}{4}$ and $I(n) = n + 1 + V(n) = n + 1 + \frac{n^2 - n\%2}{4}$.

Remark 2. If P is a *parametric polynomial* of degree n, then $[P]_{\mathcal{S}_n}$ is an $n + 1$-tuple of parameters which are treated by the constraint solving system which obtains the parameters of the basis $\mathcal{P}_n$ in the same way (see Remark 1).

4   Quantitative Analysis

In constraint solving, the number of variables occurring in the whole set of constraints usually dominates the *temporal cost* to reach a solution. In our setting, assuming P of degree n, for each representation method $V(n)$ and $I(n)$ (see Section 2) are as follows:

Method:	Hilbert	P&S	K&S	Gram	Vector
$V(n)$:	$2n+2$	$2n+2$	$n+1$	$\frac{(n+1)(n+2)}{2}$	$\frac{n^2-n\%2}{4}$
$I(n)$:	$2n+1$	$n+1$	$2n+1$	$3n+2$	$n+1+\frac{n^2-n\%2}{4}$

This table suggests the following conclusion: *for proving $Psd([0,+\infty))$, Karlin & Studden is the best choice*. However, this does *not* pay attention to the subsequent *constraint solving* process that we need to use in any implementation. In [5] an efficient procedure to solve polynomial constraints C (e.g., $P \geq 0$, where P is written as a sum of monomials with the corresponding coefficients) is given. The procedure *transforms* a polynomial constraint into a formula of the *linear arithmetic* and then fast, highly efficient *Satisfiability Modulo Theories* (SMT) techniques are used to find a solution. In linear arithmetic (logic) only constants c or additions of linear expressions $c \cdot v$ are allowed, the *atoms* consist of expressions $\ell \bowtie \ell'$ where ℓ, ℓ' are constants or linear expressions and $\bowtie \in \{=, >, \geq\}$, and the *formulas* are combinations of atoms using $\rightarrow$ (implication) and $\wedge$ (conjunction). An initial *preprocessing* $L0$ transforms $P \bowtie 0$ into $\ell_P \bowtie 0$, where ℓ_P is obtained from P by replacing the nonlinear monomials M by new variables x_M; then new atoms $x_M = M$ are added and they are subsequently transformed after further linearization using the following rules, where D is a finite domain of numbers[2]:

Definition 1. *Let C be a pure non-linear constraint and D be a finite set. The transformation rules are the following (where v is a variable):*

L1: $C \wedge x = v^p \Longrightarrow C \wedge \bigwedge_{a \in D}(v = a \rightarrow x = a^p)$, *if $p > 1$*
L2: $C \wedge x = v^p \cdot w \Longrightarrow C \wedge \bigwedge_{a \in D}(v = a \rightarrow x = a^p \cdot w)$
L3: $C \wedge x = v^p \cdot M \Longrightarrow C \wedge \bigwedge_{a \in D}(v = a \rightarrow x = a^p \cdot x_M) \wedge x_M = M$
    if M is not linear and v does not occur in M

For $x = M_0$ where M_0 is a monomial with m different variables, if M_0 consists of at most two variables, one of them of degree 1, then $L1$ or $L2$ apply; no new variables are introduced and the equality is transformed into $|D|$ new linear formulas. If $M_0 = v^p M$ contains m variables and M is not linear, then only $L3$ applies, and then introduces a *new* variable x_M together with $|D|$ new linear formulas and a new equality $x_M = M$ where M has $m-1$ variables.

Example 6. For instance, for $1 \geq c_1^2 + c_2^2$ in Example 3,

$$1 \geq c_1^2 + c_2^2 \leadsto_{L0} 1 \geq x_{c_1^2} + x_{c_2^2} \wedge x_{c_1^2} = c_1^2 \wedge x_{c_2^2} = c_2^2$$
$$\leadsto_{L1} 1 \geq x_{c_1^2} + x_{c_2^2} \wedge \bigwedge_{d \in D} c_1 = d \rightarrow x_{c_1^2} = d^2 \wedge \bigwedge_{d \in D} c_2 = d \rightarrow x_{c_2^2} = d^2$$

we obtain $1 + 2|D|$ linear formulas and 2 new variables are required.

In the following, $V_L(n)$ is the number of new variables introduced by $L0$. And if P is the targeted polynomial, p_i for $0 \leq i \leq n$ is the coefficient of x^i in P.

---

[2] Simplified definition which only uses a *single* domain of values for all variables.

Hilbert. If $f = \sum_{j=0}^{d} f_j x^j$ is a parametric polynomial of degree $d > 0$, then the coefficient c_i of x^i in f^2 is obtained from the products $f_r f_s$ such that $r + s = i$. Here, $f_s f_r$ does *not* count as a new combination because $f_r f_s + f_s f_r = 2 f_r f_s$. If $i \le d$ we have different contributing combinations from $(0, i)$ to $(i \div 2, i - i \div 2)$, i.e., $1 + i \div 2$ combinations. If $i > d$, then we have different contributing combinations starting from $(d - i, d)$, i.e., $1 + (2d - i) \div 2 = 1 + d - i \div 2 - i\%2$ combinations. Overall, if $\mu_d(c_i) = 1 + i \div 2$, if $i \le d$, and $\mu_d(c_i) = 1 + d - i \div 2 - i\%2$, if $i > d$, then c_i consists of a sum of $\mu_d(c_i)$ monomials $f_r f_s$ all of them of degree 2.

When matching $P(x) = \sum_{i=0}^{n} p_i x^i$ against Hilbert's representation, each p_i, $0 \le i \le n$ is matched by a sum c_{2i} of $2\mu_n(c_{2i})$ expressions of degree 2 (in the parameters). However, for all $0 \le i < n$, there are *additional* equations $c_{2i+1} = 0$ which are due to the *duplication* of the degree of P before the matching. Therefore, there are $2n + 1$ equations gathering

$$\sum_{i=0}^{n} 2\mu_n(c_{2i}) + \sum_{i=0}^{n-1} 2\mu_n(c_{2i+1}) = 2\left(\sum_{i=0}^{n} \mu_n(c_{2i}) + \sum_{i=0}^{n-1} \mu_n(c_{2i+1})\right)$$

quadratic terms all together, i.e., $V_L(n) = 2\left(\sum_{i=0}^{n} \mu_n(c_{2i}) + \sum_{i=0}^{n-1} \mu_n(c_{2i+1})\right)$.

Polya and Szegö. When matching $P = \sum_{i=0}^{n} p_i x^i$ against Polya and Szegö's representation in Section 2.2, if $n = 1$, then p_0 and p_1 are matched to *squared* constants f_0^2 and g_0^2, respectively. If $n = 2$, then p_1 is matched to a sum of *two* monomials of degree 2 each; finally, if $n \ge 3$, then p_0 and p_n are each of them matched to a sum of 2 squares, and each p_i, $0 < i < n$ is matched to a sum of $2\mu_{n \div 2}(c_i) + 2\mu_{(n-1) \div 2}(c_{i-1})$ expressions which are parametric coefficients: the coefficients of monomials of degree i from f_1^2 and f_2^2, and the coefficients of monomials of degree $i - 1$ from g_1^2 and g_2^2. All these parametric coefficients have degree 2. We have *two* equations with *two* terms and $n - 1$ equations gathering

$$\sum_{i=1}^{n-1} 2\mu_{n \div 2}(c_i) + 2\mu_{(n-1) \div 2}(c_{i-1}) = 2\left(\sum_{i=1}^{n-1} \mu_{n \div 2}(c_i) + \sum_{i=1}^{n-1} \mu_{(n-1) \div 2}(c_{i-1})\right)$$
$$= 2\left(1 + \mu_{n \div 2}(c_{n-1}) + \sum_{i=1}^{n-2} \mu_{n \div 2}(c_i) + \mu_{(n-1) \div 2}(c_i)\right)$$

terms. Terms M of degree 2 require a new variable x_M in the initial step $L0$. Overall, $V_L(1) = 2$, $V_L(2) = 3 \cdot 2 = 6$ and, for $n \ge 3$:

$$V_L(n) = 6 + 2\left(\mu_{n \div 2}(c_{n-1}) + \sum_{i=1}^{n-2} \mu_{n \div 2}(c_i) + \mu_{(n-1) \div 2}(c_i)\right)$$

Karlin and Studden. If $\alpha \in \{0, \dots, n\}^m$, we let $|\alpha| = \sum_{i=1}^{m} \alpha_i$. Note that $\left(\prod_{i=1}^{m}(x - a_i)\right)^n = \sum_{i=0}^{mn}(-1)^i\left(\sum_{\alpha \in \{0,\dots,n\}^m, |\alpha|=mn-i} a^\alpha\right)x^i$. If $n = 1$, there are $\binom{m}{m-i} = \binom{m}{i}$ *parametric* monomials a^α (all of them of degree $m - i$ with respect to parameters a_i) accompanying x^i. If $n = 2$, we can obtain the number of monomials accompanying x^i as follows. There are $\binom{m}{p}$ monomials a^α with

$\alpha \in \{0,1\}^m$ and $|\alpha| = m - p$. Here, $0 \leq p \leq m$. These monomials can contribute to a monomial of degree $2m - i$ for x^i. However, note that only those monomials satisfying $m - p \leq 2m - i$ (i.e., $p \geq i - m$) will be useful; otherwise, the monomials a^α *exceed* the required degree $2m - i$ for x^i. If we replace $2m - i - (m - p) = m - i + p$ occurrences of 1 by 2 in α to yield α' (with $m - p - (m - i + p) = i - 2p$ occurrences of 1 only), then, $|\alpha'| = 2(m - i + p) + i - 2p = 2m - i$ as desired. We can do that in $\binom{m - p}{m - i + p}$ different ways. However, this process makes sense only if α has *enough* occurrences of 1, i.e., if $2(m - p) \geq 2m - i$ (equivalently, $2p \leq i$, i.e., $p \leq i \div 2$) so that the replacement of occurrences of 1 by 2 in α actually leads to the appropriate α'. Overall, x^i comes with a parametric coefficient of

$$\mathsf{mon}(m, i) = \sum_{p = max(0, i - m)}^{i \div 2} \binom{m}{p} \binom{m - p}{m - i + p}$$

monomials of degree $2m - i$ (in the parameters a_i).

When matching a polynomial P of degree $2m$ against Karlin & Studden representation, we get $2m + 1$ constraints $C_i \leq p_i$, $0 \leq i \leq 2m$, where C_i consists of $\mathsf{mon}(m, i)$ monomials of degree $2m - i$ (coming from the first term of $P_{2m}(X)$ in Theorem 1) and $\mathsf{mon}(m - 1, i - 1)$ monomials of degree $2m - i$ (due to the product with β and X) coming from the second term of $P_{2m}(X)$. Therefore, C_i consists of nonlinear monomials if $2m - i > 1$ (i.e., $i < 2m - 1$). Overall, we have $\sum_{i=0}^{2m-2}(\mathsf{mon}(m, i) + \mathsf{mon}(m - 1, i - 1))$ nonlinear monomials. Similarly, P of degree $2m + 1$ yields $2m + 2$ constraints $C_i = p_i$, $0 \leq i \leq 2m + 1$, where C_i consists of $\mathsf{mon}(m, i - 1)$ monomials of degree $2m - i + 1$ (coming from the first term of $P_{2m}(X)$ above) and $\mathsf{mon}(m, i)$ monomials of degree $2m - i + 1$ (due to the product with β) coming from the second term of $P_{2m}(X)$. Therefore, C_i consists of nonlinear monomials if $2m - i + 1 > 1$ (i.e., $i < 2m$). Overall, $\sum_{i=0}^{2m-1}(\mathsf{mon}(m, i - 1) + \mathsf{mon}(m, i))$ nonlinear monomials. Hence,

$$V_L(n) = \begin{cases} \sum_{i=0}^{2m-2}(\mathsf{mon}(m, i) + \mathsf{mon}(m - 1, i - 1)) & \text{if } n = 2m \\ \sum_{i=0}^{2m-1}(\mathsf{mon}(m, i - 1) + \mathsf{mon}(m, i)) & \text{if } n = 2m + 1 \end{cases}$$

Vector. In the following, $\mu(e)$ is the *number of monomials* in a *parametric* polynomial expression e in normal form; $\kappa(e)$ is the number of *constant* monomials in e ($\kappa(e) \in \{0, 1\}$); $\lambda(e)$ is the number of *linear and non constant monomials* in e ($\lambda(e) \in \{0, 1\}$); and $\overline{\lambda}(e)$ is the number of *nonlinear* monomials in e. Clearly, $\mu(e) = \kappa(e) + \lambda(e) + \overline{\lambda}(e)$. Note that, since κ, λ, and $\overline{\lambda}$ are mutually exclusive, identifying $\mu(e)$ with one of them implies that the other are null. Finally, $\delta(e)$ is the *common* degree of all monomials in e (or $\perp$ if it does not exist). A polynomial $P_n(x)$ consists of parametric coefficients $\pi_{n,i}$ for $0 \leq i \leq n$, where $\pi_{n,n} = 1$ (i.e., $\mu(\pi_{n,n}) = \kappa(\pi_{n,n}) = 1$ and $\delta(\pi_{n,n}) = 0$). If $n > 0$ is *even* ($n = 0$ is a particular case of the previous one), then for all $0 \leq i < n$, $\pi_{n,i}$ consists of a sum of $\mu(\pi_{n,i}) = \mathsf{mon}(n \div 2, i)$ monomials, all of them of degree $n - i$ (i.e., $\delta(\pi_{n,i}) = n - i$). Thus, $\pi_{n,i}$ is linear (and nonconstant) if $n - i = 1$. Therefore, $\mu(\pi_{n,n-1}) = \lambda(\pi_{n,n-1})$ and, for all $0 \leq i < n - 1$, $\mu(\pi_{n,i}) = \overline{\lambda}(\pi_{n,i})$ and

$\delta(\pi_{n,i}) = n - i$. If n is *odd*, then $\pi_{n,0} = 0$ and for all $0 < i < n$, $\pi_{n,i}$ consists of a sum of $\mu(\pi_{n,i}) = \text{mon}(n \div 2, i - 1)$ monomials, all of them of degree $n - i + 1$ (i.e., $\delta(\pi_{n,i}) = n - i + 1$). Summarizing: $\mu(\pi_{n,i}) = \text{mon}(n \div 2, i - (n\%2))$. A constraint $P \geq 0$ is translated into a set of $n + 1$ inequalities $C_i \geq 0$, where C_i is the result of multiplying the i-th row of $M_n = (m_{ij}^n)_{n+1 \times n+1}$ and $[P]_{\mathcal{S}_n}$, the vector of coefficients of P, for $i = 0, \ldots, n$. We have the following results.

Proposition 5. *For all* n, $\mu(m_{1,2}^n) = 0$ *and for all* $1 \leq j < i \leq n$, $\mu(m_{i,i}^n) = 1$ *and* $\mu(m_{i,j}^n) = 0$. *Let* $n > 1$. *For all* $1 \leq i \leq n$,

1. $\mu(m_{i,n+1}^n) = \sum_{j=1}^{n} \mu(m_{ij}^{n-1})\mu(\pi_{n,j-1}) = \sum_{j=1}^{n} \mu(m_{ij}^{n-1})\text{mon}(n \div 2, (j-1) - n\%2)$.
2. $\delta(m_{i,n+1}^n) = \delta(m_{i,n}^{n-1}) + 1 = n + 1 - i$.

Proposition 6. $V_L(0) = V_L(1) = 0$ *and for all* $n > 1$, $V_L(n) = V_L(n - 1) + \sum_{i=1}^{n-1} \mu(m_{i,n+1}^n)$.

4.1 Comparison

Let $V_P(n) = V(n) + V_L(n)$ be the number of parameters obtained *after matching a given representation and issuing the preprocessing step L0 for the linearization*. The following table shows $V_P(n)$ for some degrees n of the targeted polynomial P for the considered representation methods[3].

Method	1	2	3	4	5	6	7	8	9	10	20	100
Hilbert	10	18	28	40	54	70	88	108	130	154	504	10504
P&S	6	10	20	28	36	46	56	68	80	94	284	5404
K&S	2	4	7	13	20	38	57	111	166	328	78741	$9.57 \cdot 10^{23}$
Vector	0	2	6	28	96	498	2322	15308	93696	758086	$2.48 \cdot 10^{16}$	$< \infty$

Although the range of values for n is small, the trend for the different methods is clear and suggests that, for $n > 6$, Pólya & Szegö's representation provides the best starting point for an implementation. Let's reason that this is actually the case. Let $W_L(n)$ be the number of variables introduced by the *linearization* after using L0 and L1, ..., L3. Obviously, $V_L(n) \leq W_L(n)$. Let $V_T(n) = V(n) + W_L(n)$ be the number of variables occurring in the linear formula obtained by the linearization process. The number $F_L(n)$ of new formulas introduced by the linearization is bounded by $|D|W_L(n) \leq F_L(n)$. And the total number of formulas is $F_T(n) = I(n) + F_L(n)$, thus bounded by $I(n) + |D|W_L(n) \leq F_T(n)$.

Since the degree of *all monomials* in the parametric polynomials in the representation is 2, for Pólya and Szegö's representation $W_L^{PS}(n) = V_L^{PS}(n)$ (the linearization process will *not* introduce more variables after L0). Thus, $V_T^{PS}(n) = V^{PS}(n) + V_L^{PS}(n) = V_P^{PS}(n)$. The $V_L^{PS}(n)$ equations $x_M = M$ are transformed by the *application* of L1 or L2 only (because $\deg(M) = 2$) into $F_L^{PS}(n) = |D|V_L^{PS}(n)$ new linear formulas. Thus, $F_T^{PS}(n) = I^{PS}(n) + |D|V_L^{PS}(n)$.

---

[3] Obtained using Haskell encodings of the cost formulas in Appendix B.

Since for $M \in \{Hilbert, KS, Vector, G\}$, $V_T^{PS}(n) = V_P^{PS}(n) < V_P^M \leq V_T^M(n)$ for all $n > 6$ (see the table above[4]), and, since $I^{PS}(n) < I^M(n)$ for all $n > 1$, we have $F_T^{PS}(n) = I^{PS}(n) + |D|V_L^{PS}(n) < I^M(n) + |D|W_L^M(n) \leq F_T^M(n)$ for all $n > 6$, we finally conclude that *Pólya and Szegö's representation is the best choice for an implementation* using the constraint solving method in [5]: it *minimizes* both the number of variables $V_T(n)$ and formulas $F_T(n)$ to be considered.

5   Related Work and Conclusions

In Section 3, we have shown that the notions of *polynomial bases* and *vector coordinates* can be used instead of that of *monomials* and *monomial coefficients* when testing univariate polynomials P for $Psd([0, +\infty))$ and $Pd([0, +\infty))$. The quantitative analysis in the previous section, though, suggests that this new method is hardly useful in practice. We show its theoretical interest as improving on the use of *Bernstein's polynomials* [3], which inspired our developments.

$Psd([0, +\infty))$ and $Psd([-1, 1])$ are related through *Goursat transform* (see [14]): Given $P \in \mathbb{R}[X]$ of degree n, we let $\widetilde{P}(X) = (1+X)^n P(\frac{1-X}{1+X})$. Furthermore, $\widetilde{\widetilde{P}}(X) = 2^n P(X)$. Then, P is $Psd([-1, 1])$ if and only if $\widetilde{P}$ is $Psd([0, +\infty))$ and $\deg(\widetilde{P}) \leq n$, see [14, Lemma 1]. Testing $Pd([-1, 1])$ or $Psd([-1, 1])$ of univariate polynomials $P \in \mathbb{R}[X]$ on $[-1, 1]$ can be done by using the so-called Bernstein's basis [6]: if $[P]_{\mathcal{B}_n} > \mathbf{0}$, for the Bernstein basis $\mathcal{B}_n$ (which consists of polynomials of degree n only) then P is $Pd([-1, 1])$ [2]. Unfortunately, $\mathcal{B}_n$ does *not* capture all $P \in Pd([-1, 1])$ as positive vectors $[P]_{\mathcal{B}_n}$. For instance, $P(X) = 5X^2 - 4X + 1$ is positive on $[-1, 1]$ but $[P]_{\mathcal{B}_2} \not> \mathbf{0}$ [6]. Nevertheless, for each $P \in Pd([-1, 1])$ of degree n the so-called *Bernstein's Theorem* [4] ensures the existence of some $p \geq n$ such that $[P]_{\mathcal{B}_p}$ consists of *positive coordinates* only (the minimum of those p is called the *Bernstein degree* of P). Unfortunately, such p can be much higher than n. For instance, for $P(X) = 5X^2 - 4X + 1$) we need to consider 23 polynomials in Bernstein's basis. Even worst, the Bernstein degree of a polynomial P is not usually known, and we have to (over)estimate it. For instance, a the recent estimation [6] is $\frac{n(n-1)}{2} \frac{M}{\lambda}$, where n is the degree of the polynomial, M is the maximum value of the coordinates $[P]_{\mathcal{B}_n}$ of P in the Bernstein basis of degree n, and λ is the minimum of P on $[-1, 1]$. For $P(X) = 5X^2 - 4X + 1$ we have $n = 2$, $M = 10$, $\lambda = \frac{1}{5}$, and a estimation of 50, far beyond 23, the real Bernstein degree of P. In [6], this problem is addressed by using *partitions* of $[-1, 1]$ where we are able to represent P in a Bernstein basis of degree n by using positive coordinates only. However, we need to produce several (up to $n + 1$) partitions of $[-1, 1]$, compute the corresponding representations of P, etc. Furthermore, it is unclear how [6] would be used with parametric polynomials (see Remark 1).

---

[4] Although we do not provide information about $V_L^G(n)$, note that $V^G(n)$ and V_T^{PS} are already very similar. Thus, assuming $V_T^{PS}(n) < V_T^G(n)$ is natural.

Example 7. For our running example, we get $\widetilde{Q}(X) = -10X^3 + 4X^2 + 10X + 4$. According to [6, page 640], for $\mathcal{B}_3 = \{ \binom{3}{i} \frac{(1-X)^{3-i}(X+1)^i}{8} \mid 0 \le i \le 3 \}$, i.e.,

$$\{ \frac{1}{8}(1 - 3x + 3x^2 - x^3), \frac{3}{8}(1 - x - x^2 + x^3), \frac{3}{8}(1 + x - x^2 - x^3), \frac{1}{8}(1 + 3x + 3x^2 + x^3) \}$$

we have: $S_{\mathcal{S}_3 \mapsto \mathcal{B}_3} = \begin{pmatrix} 1 & -1 & 1 & -1 \\ 1 & -\frac{1}{3} & -\frac{1}{3} & 1 \\ 1 & \frac{1}{3} & -\frac{1}{3} & -1 \\ 1 & 1 & 1 & 1 \end{pmatrix}$ and $[\widetilde{Q}]_{\mathcal{B}_3} = S_{\mathcal{S}_3 \mapsto \mathcal{B}_3} [\widetilde{Q}]_{\mathcal{S}_3} = \begin{pmatrix} 8 \\ -\frac{32}{3} \\ 16 \\ 8 \end{pmatrix}$,

which does *not* witness $\widetilde{Q}$ as $Psd([-1,1])$ due to the negative coordinate $-\frac{32}{3}$ in $[\widetilde{Q}]_{\mathcal{B}_3}$. The estimated Bernstein degree (for $n = 3$, $M = 16$ and $\lambda \simeq 1.22$) is 40, i.e, a 40-square cb-matrix is required! This can be compared with Example 5.

We have investigated methods for proving univariate polynomials $Psd([0, +\infty))$, and a quantitative evaluation of the requirements needed to make a practical use of them suggests that an early result by Pólya and Szegö's provides an appropriate basis for implementations in most cases. An important motivation and contribution of this work in connection with the development of tools for automatically proving termination is that we avoid the need of explicitly requiring that parametric polynomials arising in proofs of termination have non-negative coefficients (which is the usual practice in termination provers, see [8,12]). We will use our new findings in future versions of the tool MU-TERM [1].

Acknowledgements. I thank the anonymous referees for their valuable comments.

References

1. Alarcón, B., Gutiérrez, R., Lucas, S., Navarro-Marset, R.: Proving Termination Properties with MU-TERM. In: Johnson, M., Pavlovic, D. (eds.) AMAST 2010. LNCS, vol. 6486, pp. 201–208. Springer, Heidelberg (2011)
2. Basu, S., Pollack, R., Roy, M.-F.: Algorithms in Real Algebraic Geometry. Springer, Berlin (2006)
3. Bernstein, S.: Démonstration du théorème de Weierstrass fondée sur le calcul des probabilités. Communic. Soc. Math. de Kharkow 13(2), 1–2 (1912)
4. Bernstein, S.: Sur la répresentation des polynômes positifs. Communic. Soc. Math. de Kharkow 14(2), 227–228 (1915)
5. Borralleras, C., Lucas, S., Oliveras, A., Rodríguez, E., Rubio, A.: SAT Modulo Linear Arithmetic for Solving Polynomial Constraints. Journal of Automated Reasoning 48, 107–131 (2012)
6. Boudaoud, F., Caruso, F., Roy, M.-F.: Certificates of Positivity in the Bernstein Basis. Discrete Computational Geometry 39, 639–655 (2008)
7. Choi, M.D., Lam, T.Y., Reznick, B.: Sums of squares of real polynomials. In: Proc. of the Symposium on Pure Mathematics, vol. 4, pp. 103–126. American Mathematical Society (1995)

8. Contejean, E., Marché, C., Tomás, A.-P., Urbain, X.: Mechanically proving termination using polynomial interpretations. Journal of Automated Reasoning 32(4), 315–355 (2006)
9. Hilbert, D.: Über die Darstellung definiter Formen als Summe von Formenquadraten. Mathematische Annalen 32, 342–350 (1888)
10. Hong, H., Jakuš, D.: Testing Positiveness of Polynomials. Journal of Automated Reasoning 21, 23–38 (1998)
11. Karlin, S., Studden, W.J.: Tchebycheff systems: with applications in analysis and statistics. Interscience, New York (1966)
12. Lucas, S.: Polynomials over the reals in proofs of termination: from theory to practice. RAIRO Theoretical Informatics and Applications 39(3), 547–586 (2005)
13. Polya, G., Szegö, G.: Problems and Theorems in Analysis II. Springer (1976)
14. Powers, V., Reznick, B.: Polynomials that are positive on an interval. Transactions of the AMS 352(10), 4677–4692 (2000)
15. Powers, V., Wörmann, T.: An algorithm for sums of squares of real polynomials. Journal of Pure and Applied Algebra 127, 99–104 (1998)

A Rule–Based Expert System for Vaginal Cytology Diagnosis

Carlos Gamallo-Chicano[1], Eugenio Roanes-Lozano[2],
and Carlos Gamallo-Amat[3]

[1] E.T.S.I. de Telecomunicación, Universidad Politécnica de Madrid,
Avenida Complutense 30, 28040–Madrid, Spain
carlos.gamallo@gmail.com
[2] Instituto de Matemática Interdiciplinar (IMI) &
Depto. de Álgebra, Fac. de Educación, Universidad Complutense de Madrid,
c/ Rector Royo Villanova s/n, 28040–Madrid, Spain
eroanes@mat.ucm.es
http://www.ucm.es/info/secdealg/ERL/
[3] (Ret.) Depto. de Anatomía Patológica, Fac. de Medicina,
Universidad Autónoma de Madrid,
Ciudad Universitaria de Cantoblanco, 28049–Madrid
& (Ret.) Depto. de Anatomía Patológica, Hospital de la Princesa,
c/ Diego de León, 62, 28005–Madrid
cgamallo@iib.uam.es

Abstract. Vaginal cytologies are usually first checked by an expert medical laboratory technician, that is specifically dedicated to this sort of diagnosis. The cytology is only redirected to the pathologist if the diagnosis is not clear. Training these medical laboratory technicians is a long process. They have to observe many microphotographs and many real cytologies on the microscope till the required expertise is achieved. Developing this work was suggested by the third author, a just retired (2014) anatomo–pathology Professor, that suggested the possibility to somehow store his long experience working as pathologist and teacher of anatomo–pathology. The first two authors considered that it was possible to develop a Rule–Based Expert System (RBES) that tried to synthesize this knowledge and could be used as a trainer, as well as an aid for decision making by medical laboratory technician specialized in vaginal cytology. The final version will include microphotographs of real cases as illustration of all steps. The first step was to organize the knowledge in the form of a flow–chart. Once the flow–chart was constructed we thought about how to navigate through it and decided that it would be a good idea to develop a RBES that guided the user along the complex flow diagram. We decided to implement it using an algebraic inference engine (that used Groebner bases) because of our experience with this approach. We did choose the computer algebra system *CoCoA* for the implementation.

Keywords: Vaginal Cytology, Pathology, Rule Based Expert System, Groebner Basis, Computer Algebra.

G.A. Aranda-Corral et al. (Eds.): AISC 2014, LNAI 8884, pp. 34–48, 2014.
© Springer International Publishing Switzerland 2014

1   Introduction

This work arises from a comment of the third author, a just retired (2014) pathology Professor. He was sorry that his knowledge regarding evaluating cytologies, based in his long experience working as pathologist, was going to be lost.

The usual process regarding vaginal cytologies evaluation [2,17] is the following: vaginal cytologies are usually first checked by an expert medical laboratory technician, that is specifically dedicated to this sort of diagnosis. The cytology is redirected to the pathologist only if the diagnosis is not clear.

Unfortunately, training these medical laboratory technicians is a long and expensive process. They have to observe many microphotographs and many real cytologies on the microscope until the required expertise is achieved.

Moreover, the flow–chart of the diagnosis process, developed *ad hoc* for this work is complex (Figure 1). This flow–chart is inspired by the widely adopted *Bethesda System* [4,16] and somehow details the steps usually carried out in medical laboratories.

As the logic underlying the flow-chart (both the variables and the logic deductions) are Boolean, the first two authors considered that it was very well suited for approaching it as a classic Rule–Based Expert System (RBES) that synthesized this knowledge.

It could be used in two ways by medical laboratory technicians specialized in vaginal cytology:

− as a trainer, during their educational period,
− as an aid for decision making during the exercise of their profession.

The final version will include microphotographs of real cases as illustration of all steps.

We decided to implement the RBES using an algebraic inference engine (that used Groebner bases [5]) because of our experience using this approach with medical applications [10,11,12,15]. We did choose the computer algebra system *CoCoA* [1] for the implementation. The novelty of this work is not on the theoretical side but on the new application identified and the GUI under development, that will handle microphotographs as illustrations (see Section 6).

The first step carried out was to organize the knowledge in the form of a flow–chart (Figure 1).

Essentially, we have considered 7 possible diagnosis:

− malignant,
− epidermoid invasive non–keratinizing carcinoma,
− epidermoid invasive keratinizing carcinoma,
− probably malignant or slight displasia,
− perform new cytologies,
− reactive process,
− normal cytology.

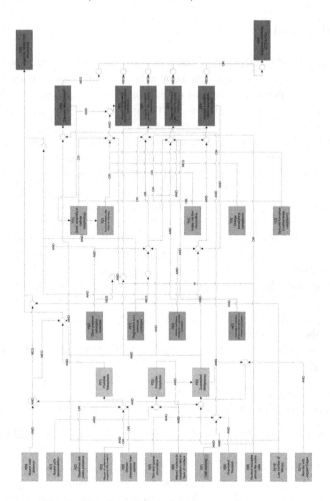

Fig. 1. The flow–chart of the diagnosis process

As a first step, we have considered 12 possibilities regarding:

- the presence or accumulation of round, polygonal or squamous cells,
- the chromatine density,
- the ratio between the nucleus and the cytoplasm,
- the existence of cell stacking,
- the size of the nucleoli,
- the number of mitosis,
- the irregularity of the border of the nuclei.

(see Figures 2–6). Other possibilities arise when advancing in the flow diagram.

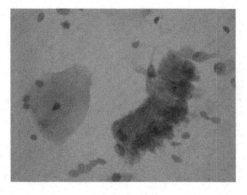

Fig. 2. Normal endocervical and squamous cells

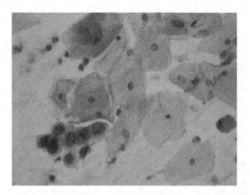

Fig. 3. Normal endocervical squamous epithelium with groups of dysplasic cells that fits in a high–grade SIL (Squamous Intraepithelial Lesion, Bethesda classification)

2   RBES Structure

Due to the characteristics of the flow–chart summarizing the diagnosis process (there is no imprecise knowledge or fuzzyness), the knowledge is structured as a simple set of (Boolean) rules.

2.1   Logical Variables and Process

The first level potential facts are $x_0, ..., x_{11}$:

- x_0: round cells presence,
- x_1: round cells accumulation,
- x_2: squamous cell plates presence,
- x_3: plates of polygonal and cohesive cells,
- x_4: denser chromatine than normal,

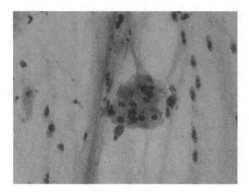

Fig. 4. Low–grade SIL (Squamous Intraepithelial Lesion, Bethesda classification). Cellular group with a low–moderate dysplasia. The cytoplasm of the cells are clearly orangephilic.

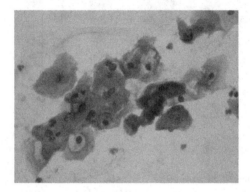

Fig. 5. Low–grade SIL (Squamous Intraepithelial Lesion, Bethesda classification). Cytopathological action of HPV (human-papillomavirus).

- x_5: thin and sparse chromatine,
- x_6: altered nucleus to cytoplasm ratio in favor of the nucleus,
- x_7: cell stacking,
- x_8: prominent nucleoli,
- x_9: nucleoli occupy almost the entire cells,
- x_{10}: low number of mitosis,
- x_{11}: nuclei with jagged edges.

From these potential facts a first level of conclusions can be obtained:

- p_1: possible metaplasia,
- p_2: possible displasia,
- p_3: suspected malignancy.

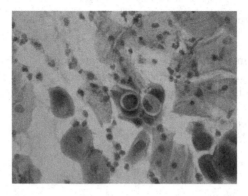

Fig. 6. Epidermoid carcinoma. Images of phagocytosis, polymorphic nuclei and hyperchromasia.

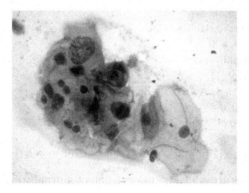

Fig. 7. Squamous cell carcinoma. Cellular groups with a clear cytologic atypia.

Now, in some cases, the system will ask about other (second level) potential facts:

- h_0: other previous cytologies available,
- h_1: regression w.r.t. previous cytologies,
- h_2: cellular membrane folding w.r.t. previous cytologies,
- h_3: chromatine evolution towards less dense w.r.t. previous cytologies.

According to the first and second level potential facts that hold and the first level conclusions, the system will ask about other (third level) potential facts:

- y_1: great variability in the cellular morphology,
- y_2: aberrant shapes (like fibers or tadpoles),
- y_3: syncytia formation (multinucleate cytoplasms),
- y_4: Indian ink–like chromatine.
- y_5: keratinizing cytoplasm.

Then a final diagnosis is reached:

- d_0: malignant,
- d_1: epidermoid invasive non–keratinizing carcinoma,
- d_2: epidermoid invasive keratinizing carcinoma,
- d_3: probably malignant or slight displasia,
- d_4: perform new cytologies,
- d_5: reactive process,
- d_6: normal cytology.

2.2 Rules

The following rules translate the processes summarized in the flow–chart:

R1: $x[2] \vee x[3] \vee x[4] \to p[1]$
R2: $x[1] \wedge x[6] \to p[2]$
R3: $p[2] \wedge x[9] \to p[3]$
R4: $\neg(x[0] \wedge x[6]) \wedge x[10] \to d[5]$
R5: $p[2] \wedge x[11] \to p[3]$
R6: $h[0] \wedge h[1] \wedge p[3] \to d[5]$
R7: $p[3] \wedge h[0] \wedge \neg h[1] \wedge \neg(h[2] \wedge h[3]) \to d[4]$
R8: $p[3] \wedge h[0] \wedge \neg h[1] \wedge h[2] \wedge h[3] \to d[3]$
R9: $p[3] \wedge \neg h[0] \to d[4]$
R10: $p[3] \wedge x[7] \wedge x[8] \wedge x[5] \to d[0]$
R11: $d[0] \wedge (y[1] \vee y[2] \vee y[3] \vee y[4] \vee x[9]) \to d[2]$
R12: $d[0] \wedge (y[1] \vee y[2] \vee y[3] \vee y[4] \vee \neg y[5] \vee x[9]) \to d[1]$

and the following rules are added in order to detect mutually excluding diagnoses:

R13: $d[0] \vee d[1] \vee d[2] \to \neg d[3] \wedge \neg d[5] \wedge \neg d[6]$
R14: $d[3] \to \neg d[0] \wedge \neg d[1] \wedge \neg d[2] \wedge \neg d[5] \wedge \neg d[6]$
R15: $d[5] \to \neg d[0] \wedge \neg d[1] \wedge \neg d[2] \wedge \neg d[3] \wedge \neg d[6]$
R16: $d[6] \to \neg d[0] \wedge \neg d[1] \wedge \neg d[2] \wedge \neg d[3] \wedge \neg d[5]$

Remark 1. It is important to underline that, if no diagnosis ($d[i]$) is obtained from these rules, then the cytology should be considered as "normal" ($d[6]$), because the flow–chart is designed to detect altered cases.

3   The Algebraic Approach to RBES

3.1   The Ring / Boolean Algebra Isomorphisms

The inference engine is based on a mathematical result, that translates the problem of determining whether a propositional formula may be inferred from others or not, into a computer algebra problem. This problem was firstly treated in [7,8] in the Boolean case and in [3,6] in the many–valued modal case. Constructing an isomorphic structure (quotient ring) allows to directly translate known results and to pass to RBES in a natural way [9,13,14].

Let $\vee, \wedge, \neg, \rightarrow$ denote the logic disjunction, conjunction, negation and implication, respectively. Let $(\mathcal{C}, \vee, \wedge, \neg, \rightarrow)$ be the Boolean algebra of the proposi-
tions that can be constructed using a finite number of propositional variables
$P, Q, ..., R$. Let us consider the Boolean algebra $(\mathcal{A}, \tilde{+}, \cdot, 1+, \text{"is a multiple"})$,
where $\mathcal{A}$ is the residue class ring

$$\mathcal{A} = \mathbb{Z}_2[p, q, ..., r]/\langle p^2 - p, q^2 - q, ..., r^2 - r \rangle$$

($\langle p^2 - p, q^2 - q, ..., r^2 - r \rangle$ denotes the polynomial ideal generated by $p^2 - p, q^2 -
q, ..., r^2 - r$). Let us define:

$$\varphi \colon (\mathcal{C}, \vee, \wedge, \neg, \rightarrow) \longrightarrow (\mathcal{A}, \tilde{+}, \cdot, 1+, \text{"is a multiple"})$$

the following way; for propositional variables:

$$P \longrightarrow p$$
$$Q \longrightarrow q$$
$$............$$
$$R \longrightarrow r$$

and for any $A, B \in \mathcal{C}$

$$A \vee B \longrightarrow a \tilde{+} b$$
$$\neg A \longrightarrow 1 + a$$

Then, as an immediate consequence of the De Morgan laws:

$$A \wedge B \longrightarrow a \cdot b$$

This correspondence turns out to be a Boolean algebra isomorphism [9,13,14].
Moreover, if $\vee$ is substituted by *xor* and $\tilde{+}$ by $+$, a (Boolean) ring isomorphism
is obtained.

The main result is Theorem 1, translating the problem of checking whether a
propositional formula is a *tautological consequence of* (i.e., *can be inferred from*)
others or not, by a polynomial ideal membership:

Theorem 1. *A propositional formula α is a tautological consequence of a set of
formulae $\{\beta_1, ..., \beta_m\}$, if and only if*

$$\varphi(\neg\alpha) \in \langle \varphi(\neg\beta_1), ..., \varphi(\neg\beta_m) \rangle$$

3.2   An Algebraic Approach to Consistency Checking and Knowledge Extraction in RBES

Bruno Buchberger developed a theory and an algorithm for finding specific basis
of ideals, which he called *Groebner basis* (named after his PhD advisor), that are
unique for each polynomial ideal [5]. In this theory, a method for computing the

Normal Form of a polynomial modulo an ideal (the residue of the polynomial modulo the ideal) is also provided. A most important application of *Normal Form* is the solution of the *ideal membership problem*: if g is a polynomial and L is an ideal:

$$g \in L \text{ if and only if NormalForm(g, L)} = 0$$

Groebner bases and Normal Forms are key for performing the simplifications derived from the definitions of the logical connectives of the logic (see Section 4.1) as well as for checking logic inferences.

Regarding RBES, logical inconsistency (i.e., what happens when a statement turns out to be true and false at the same time) is translated in the polynomial model in the degeneracy of the quotient ring into a ring with only one element (that is, a ring where $0 = 1$). This can be checked with a computer algebra system by calculating whether a certain Groebner basis is $\{1\}$ or not in the quotient ring.

Meanwhile, from Theorem 1 and the solution of the ideal membership problem mentioned above, knowledge extraction in the RBES can be performed in a computer algebra system by calculating whether the Normal Form of the polynomial translation of the negation of the logic formula belongs to the ideal generated by the polynomial translation of the negation of certain formulae (facts, rules and integrity constraints) in the quotient ring.

4 *CoCoA 4.7.5* Implementation

4.1 Defining the Polynomial Ring and the Logical Connectives (Boolean Case)

The polynomial ring is first defined:

```
A::=Z/(2)[x[0..11],p[1..3],h[0..3],y[1..5],d[0..6]];
USE A;
```

and the ideal I (that introduces idempotency) is then defined:

```
MEMORY.I:=Ideal(
  x[1]^2-x[1],x[2]^2-x[2],x[3]^2-x[3],x[4]^2-x[4],x[5]^2-x[5],
  x[6]^2-x[6],x[7]^2-x[7],x[8]^2-x[8],x[9]^2-x[9],x[10]^2-x[10],
  x[11]^2-x[11],
  p[1]^2-p[1],p[2]^2-p[2],p[3]^2-p[3],
  h[0]^2-h[0],h[1]^2-h[1],h[2]^2-h[2],h[3]^2-h[3],
  y[1]^2-y[1],y[2]^2-y[2],y[3]^2-y[3],y[4]^2-y[4],y[5]^2-y[5],
  d[0]^2-d[0],d[1]^2-d[1],d[2]^2-d[2],d[3]^2-d[3],d[4]^2-d[4],
  d[5]^2-d[5],d[6]^2-d[6]
  );
```

Then the logical connectives can be defined. Note that NEG represents ¬ and IMP represents →. As OR and AND are reserved words in *CoCoA*, we we have decided to use the (short) Spanish words: O and Y. Note that NF is the command that computes Normal Form.

```
Define NEG(M)
   Return NF(1-M,MEMORY.I);
EndDefine;

Define O(M,N)
   Return NF(M+N-M*N,MEMORY.I);
EndDefine;

Define Y(M,N)
   Return NF(M*N,MEMORY.I);
EndDefine;

Define IMP(M,N)
   Return NF(1+M+M*N,MEMORY.I);
EndDefine;
```

4.2   Defining the Rules of the RBES

```
R1:=IMP(O(O(x[2],x[3]),x[4]), p[1]);
R2:=IMP(Y(x[1],x[6]),p[2]);
R3:=IMP(Y(p[2],x[9]),p[3]);
R4:=IMP(Y(Y(NEG(Y(x[0],x[6])),p[1]),x[10]),d[5]);
R5:=IMP(Y(p[2],x[11]),p[3]);
R6:=IMP(Y(Y(h[0],h[1]),p[3]),d[5]);
R7:=IMP(Y(Y(Y(NEG(h[1]),NEG(Y(h[2],h[3]))),h[0]),p[3]),d[4]);
R8:=IMP(Y(Y(Y(NEG(h[1]),Y(h[2],h[3])),h[0]),p[3]),d[3]);
R9:=IMP(Y(p[3],NEG(h[0])),d[4]);
R10:=IMP( Y(Y(Y(p[3],x[7]),x[8]),x[5]) , d[0]);
R11:=IMP(Y(O(O(O(O(y[4],x[9]),y[3]),y[2]),y[1]),d[0]),d[2]);
R12:= IMP(Y(O(O(O(O(O(O(NEG(y[5]),x[9]),y[4]),y[3]),y[2]),y[1]),
                                              d[0]),d[1]);
R13:=IMP( O(O(d[0],d[1]),d[2]) , Y(Y(NEG(d[3]),NEG(d[5])),
                                              NEG(d[6])) );
R14:=IMP( d[3] , Y(Y(Y(Y(NEG(d[0]),NEG(d[1])),NEG(d[2])),
                                NEG(d[5])),NEG(d[6])) );
R15:=IMP( d[5] , Y(Y(Y(Y(NEG(d[0]),NEG(d[1])),NEG(d[2])),
                                NEG(d[3])),NEG(d[6])) );
R16:=IMP( d[6] , Y(Y(Y(Y(NEG(d[0]),NEG(d[1])),NEG(d[2])),
                                NEG(d[3])),NEG(d[5])) );

J:=Ideal( NEG(R1),NEG(R2),NEG(R3),NEG(R4),NEG(R5),NEG(R6),NEG(R7),
          NEG(R8),NEG(R9),NEG(R10),NEG(R11),NEG(R12),NEG(R13),
          NEG(R14),NEG(R15),NEG(R16) );
```

5   Examples

Each of the following examples are executed in less than 2 seconds in a standard computer with an i3 processor.

It has to be taken into account that the polynomials involved in this approach to consistency checking and knowledge extraction in RBES are not linear, but linear in each variable (the monomial of highest degree is the product of all the polynomial variables in the ring, and, therefore, the number of variables is an upper bound for the total degree of the polynomials) [13]. Consequently, timings are surprisingly low and problems that initially could be considered intractable due to the general double exponential worst-case complexity of Groebner bases (like a RBES involving 150 rules) are treatable in reasonable times on a standard computer.

Example 1. The facts considered are: $x[1]$, $x[6]$, $x[11]$ and $\neg h[0]$. Then we declare in *CoCoA*:

```
K:=Ideal( NEG(x[1]), NEG(x[6]), NEG(x[11]), NEG(NEG(h[0])) );
```

and we first check consistency ([1] shouldn't be obtained):

```
If GBasis(MEMORY.I+J+K)=[1]
   Then PrintLn "INCONSISTENCY"
   Else PrintLn "CONSISTENCY"
 EndIf;
```

Then we check then which of $d[0], ..., d[6]$ can be deduced (we should look for zeroes; a "1" means its negation is deduced):

```
NF(NEG(d[0]),MEMORY.I+J+K);
NF(NEG(d[1]),MEMORY.I+J+K);
NF(NEG(d[2]),MEMORY.I+J+K);
NF(NEG(d[3]),MEMORY.I+J+K);
NF(NEG(d[4]),MEMORY.I+J+K);
NF(NEG(d[5]),MEMORY.I+J+K);
NF(NEG(d[6]),MEMORY.I+J+K);
```

The output obtained is:

```
-----------------------------------
CONSISTENCY
-----------------------------------
d[0] + 1
-----------------------------------
d[1] + 1
-----------------------------------
d[2] + 1
-----------------------------------
d[3] + 1
```

```
--------------------------------
0
--------------------------------
d[5] + 1
--------------------------------
d[6] + 1
--------------------------------
```

that is, d[4] is obtained ("Perform new cytologies").

Example 2. The facts considered are: $\neg x[0]$ and $x[10]$. Then we declare in *Co-CoA*:

```
K:=Ideal( NEG(NEG(x[0])), NEG(x[10]) );
```

and we first check consistency ([1] shouldn't be obtained):

```
If GBasis(MEMORY.I+J+K)=[1]
   Then PrintLn "INCONSISTENCY"
   Else PrintLn "CONSISTENCY"
 EndIf;
```

Then we check then which of $d[0], ..., d[6]$ can be deduced (we should look for zeroes; a "1" means its negation is deduced):

```
NF(NEG(d[0]),MEMORY.I+J+K);
NF(NEG(d[1]),MEMORY.I+J+K);
NF(NEG(d[2]),MEMORY.I+J+K);
NF(NEG(d[3]),MEMORY.I+J+K);
NF(NEG(d[4]),MEMORY.I+J+K);
NF(NEG(d[5]),MEMORY.I+J+K);
NF(NEG(d[6]),MEMORY.I+J+K);
```

The output obtained is:

```
--------------------------------
1
--------------------------------
1
--------------------------------
1
--------------------------------
1
--------------------------------
d[4] + 1
--------------------------------
0
--------------------------------
1
--------------------------------
```

that is, $d[5]$ is obtained ("Reactive process"). The negations of $d[0]$, $d[1]$, $d[2]$, $d[3]$ and $d[6]$ are also obtained.

Example 3. The fact considered is: $x[0]$. Then we declare in *CoCoA*:

```
K:=Ideal( NEG(x[0]) );
```

and we first check consistency ([1] shouldn't be obtained):

```
If GBasis(MEMORY.I+J+K)=[1]
    Then PrintLn "INCONSISTENCY"
    Else PrintLn "CONSISTENCY"
 EndIf;
```

Then we check then which of $d[0], ..., d[6]$ can be deduced (we should look for zeroes; a "1" means its negation is deduced):

```
NF(NEG(d[0]),MEMORY.I+J+K);
NF(NEG(d[1]),MEMORY.I+J+K);
NF(NEG(d[2]),MEMORY.I+J+K);
NF(NEG(d[3]),MEMORY.I+J+K);
NF(NEG(d[4]),MEMORY.I+J+K);
NF(NEG(d[5]),MEMORY.I+J+K);
NF(NEG(d[6]),MEMORY.I+J+K);
```

The output obtained is:

```
-------------------------------
d[0]
-------------------------------
d[1]
-------------------------------
d[2]
-------------------------------
d[3]
-------------------------------
d[4]
-------------------------------
d[5]
-------------------------------
d[6]
-------------------------------
```

that is, no $d[i]$ is obtained. It should be considered a "Normal cytology" ($d[6]$).

6   Graphic User Interface

The *GUI* is still under development.

Using a GUI has the great advantage that the user doesn't have to deal with the CAS introducing mathematical expressions or formulae, he/she only has to click on options.

The GUI will also show the user microphotographs as illustration of the different questions and steps.

As mentioned in Section 2.1, according to the knowledge extracted, the GUI could ask the lab technician more questions in order to complete the input in order to reach the final diagnosis.

7   Conclusions

We believe that this RBES could be really useful and helpful for training lab technicians specialized in vaginal cytology and for helping them in the decision making process that takes place during their work.

This RBES is planned to be tested with new lab technicians during the 2014-15 academic year.

Acknowledgments. This work was partially supported by the research project *TIN2012-32482* (Government of Spain).

We would like to express our gratitude to Julio Rodríguez–Costa, MD, President of the *Sociedad Española de Citología (SEC)* and Head of the *Servicio de Anatomía Patológica* of the *Hospital General Universitario Gregorio Marañón* (Madrid) for providing the microphotographies in this article and for his most valuable comments.

The authors would also like to thank the anonymous reviewers for their valuable comments which helped to improve the manuscript.

References

1. Abbott, J., Bigatti, A.M., Lagorio, G.: CoCoA-5: a system for doing Computations in Commutative Algebra, http://cocoa.dima.unige.it
2. de Agustín, P.: Manual de Citología Exfoliativa Básica. Editorial Hospital Universitario 12 de Octubre, Madrid (1995)
3. Alonso, J.A., Briales, E.: Lógicas Polivalentes y Bases de Gröbner. In: Martin, C. (ed.) Actas del V Congreso de Lenguajes Naturales y Lenguajes Formales, pp. 307–315. University of Seville, Sevilla (1995)
4. American Society of Cytopathology: NCI Bethesda System, http://nih.techriver.net/index.php
5. Buchberger, B.: Bruno Buchberger's PhD thesis 1965: An algorithm for finding the basis elementals of the residue class ring of a zero dimensional polynomial ideal. Journal of Symbolic Computation 41(3-4), 475–511 (2006)

6. Chazarain, J., Riscos, A., Alonso, J.A., Briales, E.: Multivalued Logic and Gröbner Bases with Applications to Modal Logic. Journal of Symbolic Computation 11, 181–194 (1991)
7. Hsiang, J.: Refutational Theorem Proving using Term-Rewriting Systems. Artificial Intelligence 25, 255–300 (1985)
8. Kapur, D., Narendran, P.: An Equational Approach to Theorem Proving in First-Order Predicate Calculus. In: Proceedings of the 9th International Joint Conference on Artificial Intelligence (IJCAI 1985), vol. 2, pp. 1146–1153 (1985)
9. Laita, L.M., Roanes-Lozano, E., de Ledesma, L., Alonso, J.A.: A computer algebra approach to verification and deduction in many-valued knowledge systems. Soft Computing 3, 7–19 (1999)
10. Laita, L.M., Roanes–Lozano, E., Maojo, V., Roanes–Macías, E., de Ledesma, L., Laita, L.: An Expert System for Managing Medical Appropriateness Criteria based on Computer Algebra Techniques. Computers and Mathematics with Applications 42(12), 1505–1522 (2001)
11. Pérez Carretero, C., Laita, L.M., Roanes–Lozano, E., Lázaro, L., González–Cajal, J., Laita, L.: A Logic and Computer Algebra-Based Expert System for Diagnosis of Anorexia. Mathematics and Computers in Simulation 58(3), 183–202 (2002)
12. Piury, J., Laita, L.M., Roanes–Lozano, E., Hernando, A., Piury–Alonso, F.J., Gómez–Argüelles, J.M., Laita, L.: A Gröbner bases-based rule based expert system for fibromyalgia diagnosis. Revista de la Real Academia de Ciencias. Serie A. Matemáticas (RACSAM) 106(2), 443–456 (2012)
13. Roanes-Lozano, E., Laita, L.M., Hernando, A., Roanes-Macías, E.: An algebraic approach to rule based expert systems. Revista de la Real Academia de Ciencias. Serie A. Matemáticas (RACSAM) 104(1), 19–40 (2011), doi:10.5052/RACSAM.2010.04
14. Roanes-Lozano, E., Laita, L.M., Roanes-Macías, E.: A Polynomial Model for Multivalued Logics with a Touch of Algebraic Geometry and Computer Algebra. Mathematics and Computers in Simulation 45(1), 83–99 (1998)
15. Rodríguez–Solano, C., Laita, L.M., Roanes–Lozano, E., López–Corral, L., Laita, L.: A Computational System for Diagnosis of Depressive Situations. Expert Systems with Applications 31, 47–55 (2006)
16. Soloman, D.: The, Bethesda system for reporting cervical/vaginal cytologic diagnoses: Developed and approved at the national cancer institute workshop in Bethesda, MD, December 12-13, Diagnostic Cytopathology 5, 331–334 (1988), doi: 10.1002/dc.2840050318
17. Viguer, J.M., García del Moral, R.: Laboratorio y Atlas de Citología. Editorial Interamericana McGraw–Hill, Madrid (1995)

Obtaining an ACL2 Specification
from an Isabelle/HOL Theory*

Jesús Aransay-Azofra, Jose Divasón, Jónathan Heras, Laureano Lambán,
María Vico Pascual, Ángel Luis Rubio, and Julio Rubio

Departamento de Matemáticas y Computación, Universidad de La Rioja, Spain
{jesus-maria.aransay,jose.divasonm,jonathan.heras,lalamban,
mvico,arubio,julio.rubio}@unirioja.es

Abstract. In this work, we present an interoperability framework that
enables the translation of specifications (signature of functions and lemma
statements) among different theorem provers. This translation is based on
a new intermediate XML language, called XLL, and is performed almost
automatically. As a case study, we focus on porting developments from
Isabelle/HOL to ACL2. In particular, we study the transformation to
ACL2 of an Isabelle/HOL theory devoted to verify an algorithm computing a diagonal form of an integer matrix (looking for the ACL2 executability that is missed in Isabelle/HOL). Moreover, we provide a formal proof
of a fragment of the obtained ACL2 specification — this shows the suit-
ability of our approach to reuse in ACL2 a proof strategy imported from
Isabelle/HOL.

1 Introduction

In the frame of the ForMath European project [1], several theorem provers are
used to verify mathematical algorithms, with an emphasis on Coq/SSReflect [11]
but also using intensively Isabelle/HOL [19] and ACL2 [16]. Due to this diversity
of tools, it was natural to investigate how different provers could collaborate, in
some manner, in the same formalisation effort.

Numerous contributions have been made along the years in the area of theorem proving interoperability. We give here just a few strokes of the brush, by
saying that translations among proof assistants can be of two kinds: *deep* and
shallow. In the former, deep translations, e.g. [6,12,15,18], the soundness of the
transformation is ensured, and thus, it is necessary to analyse semantics issues
(underlying logics, language expressiveness, and so on). In the latter, shallow
translations, e.g. [10,17,21], only the syntactical structure is translated from the
source formalism to the target one.

In this work, and starting from a complete formalisation in Isabelle/HOL, we
develop a set of tools that translates a *proof plan* to ACL2, looking for efficient

---

* Partially supported by Ministerio de Ciencia e Innovación, project MTM2009-13842,
  by European Union's 7th Framework Programme under grant agreement nr. 243847
  (ForMath), and by Universidad de La Rioja, research grant FPI-UR-12.

G.A. Aranda-Corral et al. (Eds.): AISC 2014, LNAI 8884, pp. 49–63, 2014.

(but verified) *executability*. Since we do not have a *deep* Isabelle/HOL–ACL2 translator at our disposal, we try to materialise the previous observation by writing a *shallow* porting mechanism. Even if we do not aim at doing a survey of the state of the art in the field, it is worth noting that approaches abound in the literature. The Omega system [23] has been fruitfully used through the years to perform proof planning strategies. As far as we can determine, it is unable to integrate with the theorems provers we are interested in. A different tool is the Evidential tool bus [9]. Evidential's design principle is that of *semantic neutrality*; our work here is probably not different from it, but could be seen as an *ad-hoc* case study of what they call *translators* (in our case, from Isabelle/HOL to ACL2). Another meaningful possibility for our development would have been the use of the THF0 [4] language, to which several tools in TPTP are translated; indeed, a subset of Isabelle/HOL statements can be already translated to THF0, in order to enable the communication with external automated theorem provers. Applying a similar idea to ACL2 seems an interesting idea, but our interest in XML based tools, such as Ecore and OCL (see [3] for details), leads us to propose a different approach.

Our approach translates function signatures and statements, while proofs and function bodies are not ported. In principle, this weak process could be considered unsafe (this criticism could be also applied to any *shallow* strategy). Nevertheless, our key idea is based on the following argument: *the family of function signatures and statements in a formalisation encodes a* proof scheme *that can be reused in any other system*. Of course, some constraints must be added to render sensible this claim. For instance, the target framework must be expressive enough to receive the formulas from the source environment (at least, in the concrete problem to be ported). Additionally, such a reuse may be not optimal (otherwise, something as a *deep* translation would be accomplished), because both the data structures and the working style of each theorem prover can be very distant. In any case, at some convenient abstract level, the *sketch of the proof* can be translated, saving a significant amount of time. At the end of the process, when a complete proof is (re)built in the target system, the question about the soundness of the translation is no longer relevant.

The above proposal is instantiated in this paper in a particular case study, where we go from Isabelle/HOL to ACL2, transforming a complete constructive proof in Isabelle/HOL, related to integer matrices manipulation, into an (incomplete) ACL2 specification. This transformation is justified because it is not possible to directly execute the matrix operations inside Isabelle/HOL — due to the internal representation of matrices — and we decided to look for a proof in ACL2, where executability will be guaranteed. The essence of the proof is captured in this transformation, showing the adequacy of our contribution.

The organization of the paper is as follows. Our general framework to interoperate is briefly described in Section 2. Section 3 is devoted to comment on the Isabelle/HOL theory developed and the translation process, while Section 4 deals with the completion of the ACL2 specification until a proof of a fragment of the theory is obtained. The paper ends with conclusions, further work and the bibliography. The paper is backed with a report [3] in which we have thoroughly described the architecture of the tool, two case studies, and the steps which can be applied to produce new translations.

2 A (Minimal) Framework to Interoperate

The framework presented in [3] — from now on called *I2EA* (Isabelle/HOL to Ecore and ACL2) — allows the transformation of Isabelle/HOL specifications to both Ecore models [2] and ACL2 specifications; the role of Ecore is presented in [3]. In this paper, we only focus on proving the following concept: *the I2EA framework can be used to translate Isabelle/HOL specifications to ACL2, and to reuse a proof scheme in ACL2 imported from Isabelle/HOL.* To this aim, we just use the components of the I2EA framework shown in Figure 1. We describe the components of that diagram in the following subsections.

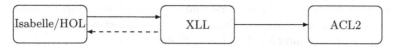

Fig. 1. (Reduced) Architecture of the I2EA framework

2.1 Isabelle

Isabelle [19] is a generic interactive proof assistant, on top of which different logics can be implemented; the most explored of this variety of logics is higher-order logic (or *HOL*), and it is also the logic where a greater number of tools (code generation, automatic proof procedures) are available.

The HOL type system is rather simple; it is based on non-empty types, function types (⇒) and type constructors κ that can be applied to already existing types (*nat, bool*) or type variables (α, β). Types can be also introduced by enumeration (*bool*) or by induction, as lists (by means of the *datatype* command). Additionally, new types can be also defined as non-empty subsets of already existing types by means of the *typedef* command; the command takes a set defined by comprehension over a given type $\{x :: \alpha . P\,x\}$, and defines a new type σ, as well as *Rep* and *Abs* morphisms between the types. *Type annotations* can be made explicit to the prover, by means of the notation $x :: \alpha$, and can solve situations where types remain ambiguous even after type inference.

Isabelle also introduces type classes in a similar fashion to Haskell; a type class is defined by a collection of operators (over a single type variable) and premises over them. For instance, the library has type classes representing arithmetic operators (like sum or unary minus). Concrete types (*int, real, set*, and so on) are proved to be *instances* of those type classes. The expression ($x :: \alpha :: plus$) imposes that the type variable α poses the structure and properties stated in the *plus* type class, and can be later replaced exclusively by types which are instances of such a type class. Type classes provide operator overloading, enabling to reuse symbols for different types (0 :: *nat* and 0 :: *int*).

2.2   ACL2

ACL2 [16] stands for "A Computational Logic for Applicative Common Lisp". Roughly speaking, ACL2 is a programming language, a logic and a theorem prover. Its programming language is an extension of an applicative subset of Common Lisp [24]. The ACL2 logic describes the programming language, with a formal syntax, axioms and rules of inference: the applicative subset of Common Lisp is a model of the ACL2 logic. Finally, the theorem prover provides support for mechanised reasoning in the logic. Thus, the system constitutes an environment in which programs can be defined and executed, and their properties can be formally specified and proved with the assistance of a theorem prover. The logic is a first-order logic with equality including axioms for propositional logic and for a number of primitive Common Lisp functions and data types.

New function definitions (using `defun`) are admitted as axioms only if there exists an ordinal measure in which the arguments of each recursive call (if any) decrease, thus proving its termination and ensuring that no inconsistencies are introduced. The operator `defun-sk` introduces new functions that represent existential quantifiers, following the idea of *Skolemization*.

The ACL2 theorem prover is an integrated system of *ad-hoc* proof techniques, including simplification and induction among them. Simplification is a process combining term rewriting with some decision procedures (linear arithmetic, type set reasoner, and so on). Sophisticated heuristics for discovering an (often suitable) induction scheme is one of the key features in ACL2. The command `defthm` starts a proof attempt, and, if it succeeds, the theorem is stored as a rule (in most cases, a conditional rewriting rule). The theorem prover is automatic in the sense that, once `defthm` is submitted, the user can no longer interact with the system. However, in some sense, it is interactive. Often, non-trivial results cannot be proved on a first attempt, and then the role of the user is important: she has to guide the prover by providing a suitable collection of definitions and lemmas, used in subsequent proofs as rewrite rules. These lemmas are suggested by a preconceived "hand" proof (at a higher level) or by inspection of failed proofs (at a lower level). This kind of interaction is called "the Method" [16].

2.3   XLL

XLL, for *Xmall Logical Language*, is an XML-based specification language. Its definition is done through an XML schema [3, Appendix 6.7] which consists of two parts:

1. A specification of *data types* (or classes), including for each data type a name plus a family of operators (or methods).
2. A set of logical statements, expressing some properties of the data types involved.

The first part defines a dictionary for the operations that can appear in the second one. In the second part, XLL defines essentially a typed first-order logic language. The propositional connectives are grouped in the first part of the XLL schema, the one referring to data types, and will be translated *literally* to any other specification language (for example, ACL2) as *primitive* operations.

With respect to the types in the logical expressions, they can be user-defined classes or elementary data types which can be easily inferred from the context. Only in cases of implicit coercion, some additional type annotations are necessary. For instance, in integer matrix manipulation, the constant 1 can denote either an entry of a matrix or an index for a row or column. In the former case, 1 should be considered as an integer; on the contrary, in the latter, it must be considered as a natural number. These disambiguation annotations are encoded inside the very logical expression, by using enriched arguments like:

```
<constant> <name>1</name> <type>Nat</type> </constant>
```

Additionally, the schema checks that the statements of the properties contain operations that exclusively appear in the XLL file itself (in the specification part); the XLL schema ensures that the properties stated in the file are referred to a certain context (a set of data types and operations).

We have not specified a formal semantics of XLL; it is a simple language in which types, operations and logical statements over them (in a typed first-order logical language) can be expressed. The language is enough to cover both the expressiveness of ACL2, and a first-order fragment of Isabelle/HOL.

In the case study presented in Section 3, XLL documents (that is to say, XML documents compliant with our XLL schema) are generated from Isabelle/HOL formalisations. As an intermediary step, we use a set of libraries generating XML documents from Isabelle specifications, part of the Isabelle standard distribution. Namely, we generate a collection of XML files from an Isabelle/HOL theory, which are subsequently transformed into an XLL file. Furthermore, from that XLL document an ACL2 set of statements can be also generated, essentially forgetting the data types part, because ACL2 is an environment without explicit static typing; nevertheless, the type annotations in the logical expressions are used to generate predicates checking dynamically ACL2 types, as we will explain later. From the XLL document, we are able to produce an Isabelle theory, and automatically prove (in Isabelle!) the behavioural equivalence between the generated Isabelle theory (from the XLL document) and the original Isabelle theory — see [3] for an example. However, it is not possible to reconstruct the Isabelle theory from the produced ACL2 specification, because, having ACL2 a weaker type system than Isabelle/HOL, we irretrievably lost information in the translation.

Each one of the previous steps is automatic, except the initial choice of the types, operations and lemmas which are of interest for our development (types and operations dependencies are also solved by the tool). The user is in charge of choosing the definitions (and lemmas) that will be exported, and she has to decide what is the correct level of granularity to export a set of functions (and lemmas) detailed-enough to be useful for the proof-scheme, but also abstract-enough to give a proof-scheme independent from the concrete representation of the source theorem prover. This is why we have labelled the whole generation process as *almost* automatic.

3   Transforming an Isabelle/HOL Formal Development to ACL2: A Diagonal Matrix Form

In this section, we apply the previously defined interoperability setting to an Isabelle/HOL formalisation of some well-known results about integer matrices. It is important to highlight that, even if the theory is written in HOL, the problem is essentially of a first-order nature, and therefore the information that is lost when going from Isabelle/HOL to XLL (and then to ACL2) does not prevent us from getting a sensible specification. Thus, we consider an Isabelle/HOL development (described in [3]) which defines a verified method to reduce a given matrix to a diagonal form, i.e. a method to compute a diagonal matrix which is *equivalent* to the initial one — two matrices A and B are equivalent if there exist two invertible matrices P and Q such that $B = PAQ$.

Then, the corresponding Isabelle/HOL formalisation includes the basic matrix operations (addition and multiplication) and properties of the ring of integer matrices. The main result of this Isabelle/HOL theory can be expressed as follows:

Lemma 1. *Given an integer matrix A, there exist three integer matrices P, Q and B such that:*

- $B = PAQ$;
- *P and Q are invertible matrices;*
- *B is a diagonal matrix.*

The diagonal matrix presented in the previous lemma is usually computed in many algorithms as an intermediary step in the computation of the Smith Normal Form (see [5, 7]); indeed, this particular matrix has its own interesting properties.[1]

3.1   An Isabelle/HOL Formalisation of Lemma 1

Let us briefly describe the Isabelle/HOL formalisation that leads us to prove Lemma 1 (the interested reader can find a more complete description in [3]).

One of the most relevant decisions in the initial steps of a formalisation is the choice of a suitable representation for the objects involved in the development; in this particular case, integer matrices. In our Isabelle/HOL theory, the family

---

[1] In spite of the fact that the calculation on the Smith Normal Form is a more renowned result than the one presented in Lemma 1, the diagonal form is enough in many calculations. For example, the homology of a chain complex over a ring can be obtained using the diagonal form of the differential maps represented as matrices. This situation is usual in some programs for Symbolic Computation in Algebraic Topology; thus, the presented result has its own interest in that area.

of matrices is represented as the set of functions with two arguments of type
nat and finitely many non-zero positions. This functional representation eases
the definition of operations over matrices and the proof of properties (it has
been introduced in Isabelle, and successfully used, as a part of the Flyspeck
project [20]). The formal definition is:

```
type_synonym 'a infmatrix = "nat => nat => 'a"

definition nonzero_positions ::
        "('a::zero) infmatrix ⇒ (nat x nat) set" where
        "nonzero_positions A = {pos. A (fst pos) (snd pos) ~= 0}"

definition "matrix = {(f::(nat ⇒ nat ⇒ 'a::zero)).
                    finite (nonzero_positions f)}"

typedef 'a matrix = "matrix :: (nat ⇒ nat ⇒ 'a::zero) set"
```

The library offers several definitions and properties over this data type; in particular, operations *nrows* and *ncols* that, making use of the Hilbert's ϵ operator,
return the maximum row and column which contain nonzero elements. We had
to define elementary operations on matrices; in particular, there are two basic
operations for our development (*interchange_rows_matrix* and *row_add_matrix*)
that exchange two rows of a matrix, and replace a row by the sum of itself
and another row multiplied by an integer respectively (the corresponding operations acting on columns are also defined). We introduce here the definition
of *interchange_rows_matrix*, as well as the definition of its *functional* behaviour
over the *underlying* representation of matrices presented previously (in this case,
functions):

```
definition interchange_rows_infmatrix ::
        "int infmatrix ⇒ nat ⇒ nat ⇒ int infmatrix"
  where "interchange_rows_infmatrix A n m ==
     (λi j. if i=n then A m j else if i=m then A n j else A i j)"

definition interchange_rows_matrix ::
        "int matrix ⇒ nat ⇒ nat ⇒ int matrix"
  where "interchange_rows_matrix A n m==
        Abs_matrix (interchange_rows_infmatrix (Rep_matrix A) n m)"
```

The previous definition relies on the type morphisms **Abs_matrix** and
Rep_matrix, which perform the conversion between the type (**matrix**) and the
underlying type (**infmatrix**, an abbreviation of the functional representation
of matrices). It makes use of the function **interchange_rows_infmatrix**, which
represents the functional behaviour of the elementary operation.

Using these functions (and their column counterparts), we can define several
auxiliary results and finally state and prove Lemma 1 in Isabelle/HOL.

```
lemma Diagonalize_theorem:
shows "∃P Q B. is_invertible P ∧ is_invertible Q ∧ B = P*A*Q
  ∧ is_square P (nrows (A::int matrix)) ∧ is_square Q (ncols A)
  ∧ Diagonalize_p B (max (nrows A) (ncols A))"
```

The proof of this result (from a conceptual point of view) is *constructive*; that is, the witnesses (P, Q and B in this case) for the existential expression (it applies the existential quantifier to P, Q and B) are explicit and could be algorithmically produced. Roughly speaking, the procedure used in the proof to compute the equivalent diagonal matrix is analogous to the Gauss-Jordan elimination method [8, Section 28.3].

The fact that the essence of the proof is constructive would allow us to obtain executable programs from it. However, it is not possible to directly execute the corresponding expressions inside Isabelle due to the representation of matrices (based on the abstract type *matrix*). It is well-known that, in such a situation, we could *refine* the data structure representation (by taking, for instance, lists of lists to represent matrices), in order to get executable code from Isabelle. Our approach is however different: we look for a proof in a different theorem prover (ACL2), where executability will be guaranteed.

3.2   From the Isabelle/HOL Theory to XLL

As a first step in the translation of the Isabelle/HOL theory to ACL2, the corresponding XLL document — an XML instance compliant with the XLL schema — is generated through a series of automatic steps, as presented in [3]. The XLL description of the theory consists of two different components: data types and logical statements.

The former contains the specification of the data types appearing in the source Isabelle/HOL theory (their names and the collection of functions where the types appear as parameters, which are selected by the tool). This information is automatically organized in a *class*, an XLL structure which is used to represent each type and its operations (and that in our Ecore experiments is later assigned to a UML class), whose XLL description (for the type `matrix`) appears in Figure 2 (we only include elementary operations over matrices).

The second component of the resulting XLL document consists of a set of statements establishing the properties of the entities (data and methods) involved in the theory. To illustrate this component, we include in Figure 3 (Page 58) the XLL description of the lemma which states in Isabelle/HOL the idempotence property of `interchange_rows_matrix` — the square of this function is the identity.

```
lemma interchange_rows_matrix_id:
 shows
  "interchange_rows_matrix (interchange_rows_matrix A n m) n m = A"
```

```
<Class name="Matrix.matrix">
  <Class_Parameters>
    <Parameter name="alpha">
      <Type name="Int.int"/>
    </Parameter>
  </Class_Parameters>
      ...
  <method name="Diagonal_form.interchange_rows_matrix">
    <Type name="Matrix.matrix"/>
    <Input name="n"><Type name="Nat.nat"/></Input>
    <Input name="m"><Type name="Nat.nat"/></Input>
  </method>
  ...
</Class>
```

Fig. 2. XLL for the generated `matrix` class

3.3 From XLL to ACL2

From the XLL description of the Isabelle/HOL theory, and only applying XSLT transformations [25], we can obtain an ACL2 specification. For instance, the ACL2 function obtained from the XLL file of Figure 2 is:

```
(defun Diagonal_form.interchange_rows_matrix (A n m)
    (declare (ignore A n m)) nil)
```

The body of this function is empty (in fact, the value nil is returned by default to allow the compilation of the function) since data types are very linked to the way of working in each proof assistant. Therefore, it is unlikely that the Isabelle/HOL representation of matrices will be the most useful one to work in ACL2. Then, we delegate the task of defining a suitable representation of the data types to a further step in the development process (see Section 4).

In the same way, theorems like the one presented in Figure 3 are also translated to ACL2.

```
(defthm interchange_rows_id
    (implies
      (and (matrix_integerp A) (natp n) (natp m))
      (equal (Diagonal_form.interchange_rows_matrix
          (Diagonal_form.interchange_rows_matrix A n m) n m) A)))
```

Using this procedure, we translate the whole Isabelle/HOL development into ACL2. The ACL2 version of Lemma 1 is stated as follows.

```
(defun-sk exists_Diagonalize_theorem (A)
  (exists (P Q B)
      (and (Diagonal_form.is_invertible P)
```

```
<Theorem>
    <name>interchange_rows_id</name>
    <forall>
      <param> <name>A</name> <type>Int.int Matrix.matrix</type> </param>
      <body>
        <forall>
          <param> <name>n</name> <type>Nat.nat</type> </param>
          <body>
            <forall>
              <param> <name>m</name> <type>Nat.nat</type> </param>
              <body>
                <operation>
                  <name>HOL.eq</name>
                  <operation>
                    <name>Diagonal_form.interchange_rows_matrix</name>
                    <operation>
                      <name>Diagonal_form.interchange_rows_matrix</name>
                      <constant> <name>A</name> </constant>
                      <constant> <name>n</name> </constant>
                      <constant> <name>m</name> </constant>
                    </operation>
                    <constant> <name>n</name> </constant>
                    <constant> <name>m</name> </constant>
                  </operation>
                  <constant> <name>A</name> </constant>
                </operation>
              </body>
            </forall>
          </body>
        </forall>
      </body>
    </forall>
</Theorem>
```

Fig. 3. XLL for the interchange_rows_id theorem

```
        (and (Diagonal_form.is_invertible Q)
        (and (equal B (Groups.times_class.times
                      (Groups.times_class.times P A) Q))
        (and (Diagonal_form.is_square P (Matrix.nrows A))
        (and (Diagonal_form.is_square Q (Matrix.ncols A))
        (Diagonal_form.Diagonalize_p B
            (max (Matrix.nrows A) (Matrix.ncols A)))))))))))

(defthm Diagonalize_theorem
  (implies (matrix_integerp A)
          (exists_Diagonalize_theorem A)))
```

We claim that the previous statements, transferred from the Isabelle/HOL formal development, can be used as a guideline to achieve a similar formalisation in ACL2. The following section includes a small example illustrating this fact.

4 An Experiment in Reusing (Schemes of) Proofs

As a driving example to translate proof schemes, we consider a small theory about the basic operations over matrices described in Section 3. In particular, we are interested in proving that the elementary matrices $P_{i,j}$ (identity matrices where the i-th and j-th rows are swapped) are invertible. The actual statement of the lemma explains that the square of a $P_{i,j}$ (encoded using the function P_ij) matrix is the identity matrix.

```
lemma P_ij_invertible:
  assumes n: "n < a" and m: "m < a"
  shows "(P_ij a n m) * (P_ij a n m) = one_matrix (a)"
```

The main components of the theory required to prove the above lemma are the functions:

- interchange_rows_matrix, that exchanges two rows of a matrix;
- P_ij, that defines the elementary matrix P_{ij} in dimension a.

```
  definition P_ij :: "nat ⇒ nat ⇒ nat ⇒ int matrix"
    where "P_ij a n m ==
            interchange_rows_matrix (one_matrix a) n m"
```

and the lemmas:

- interchange_rows_matrix_id, which states the idempotency of the function interchange_rows_matrix (see the end of Subsection 3.2).
- PA_interchange_rows, that relates the interchange_rows_matrix operation to the left product by the P_ij matrices.

```
  lemma PA_interchange_rows:
    assumes n:"n < nrows A" and m: "m < nrows A"
            and na: "nrows A <= a"
    shows "interchange_rows_matrix (A::int matrix) n m =
          (P_ij a n m) * A"
```

This specification can be considered as a suitable strategy to prove lemma P_ij_invertible in Isabelle/HOL and, as we will show, the same strategy can be replicated in ACL2 to prove the same result.

We omit the XLL instance provided by this source Isabelle/HOL theory (the interested reader can extract it from [3]) to directly present the ACL2 specification which is automatically produced by the I2EA framework. The file generated by the I2EA framework consists of two parts: the headers of the functions and the lemmas.

In the function section, we find not only the specification of the functions defined in the Isabelle/HOL theory, but also all the functions involved in the theorems of such a theory which are defined in other libraries (for instance, the definition of the identity matrix `Matrix.one_matrix`), and also predicate recognisers (in this case, the `matrix_integerp` function is the recogniser for integer matrices), which replace in ACL2 some Isabelle typing information:

```
(defun matrix_integerp (x) (declare (ignore x)) nil)

(defun Diagonal_form.interchange_rows_matrix (x1 x2 x3)
    (declare (ignore x1 x2 x3 )) nil)

(defun Matrix.nrows (x1)
    (declare (ignore x1)) nil)

(defun Groups.times_class.times (x1 x2)
    (declare (ignore x1 x2)) nil)

(defun Diagonal_form.P_ij (x1 x2 x3)
    (declare (ignore x1 x2 x3)) nil)

(defun Matrix.one_matrix (x1)
    (declare (ignore x1)) nil)
```

Using the headers of these functions as a guideline, we must provide a concrete representation for integer matrices and define the rest of the functions — we re-use an ACL2 matrix library presented in [13], where matrices are encoded as lists of vectors, and several background lemmas are provided; in addition, ACL2's pre-defined functions are used to define the body of some functions (e.g. the "*" ACL2's function is used to define the body of the function `Groups.times_class.times`).

Once this task is carried out, we can focus on the lemmas generated by the I2EA framework.

```
(defthm interchange_rows_id
  (implies (and (matrix_integerp A) (natp n) (natp m))
           (equal (Diagonal_form.interchange_rows_matrix
                     (Diagonal_form.interchange_rows_matrix A n m)
                     n m) A)))

(defthm PA_interchange_rows
  (implies (and (natp n) (matrix_integerp A) (natp m) (natp a)
                (< n (Matrix.nrows A)) (< m (Matrix.nrows A))
                (<= (Matrix.nrows  A) a))
```

```
                (equal (Diagonal_form.interchange_rows_matrix A n m)
                       (Groups.times_class.times
                                     (Diagonal_form.P_ij a n m) A))))

(defthm P_ij_invertible
   (implies (and (natp n) (natp a) (natp m) (< n a) (< m a))
            (equal (Groups.times_class.times
                          (Diagonal_form.P_ij a n m)
                          (Diagonal_form.P_ij a n m))
                   (Matrix.one_matrix a))))
```

ACL2 is able to find the proof of the first two lemmas without any external help; however, it gets stuck when proving lemma `P_ij_invertible`. We can suggest ACL2 to use the lemmas `PA_interchange_rows` and `interchange_rows_id` to finish the proof, but the system is not able to use them. Inspecting ACL2's proof attempt, we realise that ACL2 needs a lemma which states that the function `Diagonal_form.P_ij` generates an integer matrix.

```
(defthm P_ij_matrix_integerp
   (implies (and (natp a) (natp n) (natp m))
            (matrix_integerp (Diagonal_form.P_ij a n m))))
```

Once this lemma is introduced in the system, ACL2 finishes the proof of `P_ij_invertible`. Let us note that `P_ij_matrix_integerp` is taken for granted in Isabelle/HOL, since this type information is already provided in the definition of `P_ij`.

As foreseen, the previous discussion shows that we can import the Isabelle/HOL proof scheme into ACL2, but some additional lemmas can be necessary to complete the proof — in our experiments, those auxiliary lemmas are always related to predicate lemmas such as `P_ij_matrix_integerp`. These ACL2 lemmas containing the information encoded in the Isabelle functions target types, in the form of recognisers, will be automatically generated in future releases of the I2EA framework. The case study on matrices has proven itself useful to give us feedback on the kind of information that is represented differently in Isabelle and ACL2, but still necessary on both tools.

5   Conclusions and Future Work

In this paper, we have described a facility to transform Isabelle/HOL theories into ACL2 specifications. We have shown, through a concrete case study, that the transferred-information is enough to reconstruct a proof in ACL2. In particular, the original Isabelle theory consists of 5952 lines of code, 222 lemmas, and 54 definitions. Those lemmas and definitions have been filtered to extract 119 lemmas, and 32 definitions that have been translated to ACL2. Finally, the ACL2 development consists of 58 definitions (19 of them using defun-sk) and 119 lemmas.

The drawbacks of our approach (with respect to other mainstream approaches to interoperate between theorem provers) are the following ones:

– Our proposal is not *universal*, in the sense that it is not proposed as a general solution to the interoperability problem.
– Our proposal is *partial*, because when going from Isabelle/HOL to ACL2, it is evident that no higher-order Isabelle theory could be translated to a, necessarily first-order logic, ACL2 specification.
– Our proposal is *incomplete* (even for the fragment of Isabelle/HOL that we are considering), since we port only function signatures and statements, while definitions and proofs are not transferred in the process.
– Our proposal requires the expert knowledge of the user to choose the relevant lemmas and definitions to generate a useful proof-scheme.

On the positive side, the benefits of the presented framework are:

– Our proposal was developed *quickly* (at least when comparing it with the effort required to embed a system in another one); in the implementation we used many already available XML tools, reducing the programming needs to a minimum [3].
– Our proposal is *flexible*, because due to the lightweight technology used, we have been able to modify our XLL schema to adapt it to other close situations, without reprogramming the whole framework, see [14].
– And last, but not least, our proposal *works*, since we have shown how a nontrivial formalisation (a diagonalisation algorithm for integer matrices) has been translated profitably to ACL2, as required in our ForMath setting.

We think that the global balance is positive. More research and experiments are needed in order to get more evidences of the interest of this kind of *shallow* interoperability approach. As future work, we should translate other (first-order like) Isabelle/HOL theories to ACL2; for instance, it would be interesting to study algorithms for symbolic matrices presented in [22]. Moreover, we should generalise our approach to other proof assistants. In this last line, some successful experiences have been already made: we have used XLL as intermediary language to port Coq statements to ACL2 in a context of Java programming verification, see [14].

References

1. ForMath: Formalisation of Mathematics, European project, http://wiki.portal.chalmers.se/cse/pmwiki.php/ForMath/ForMath
2. MDT/OCL in Ecore, http://wiki.eclipse.org/MDT/OCLinEcore
3. Aransay, J., et al.: A report on an experiment in porting formal theories from Isabelle/HOL to Ecore and ACL2. Technical report, ForMath European project (2013), http://wiki.portal.chalmers.se/cse/uploads/ForMath/isabelle_acl2_report
4. Benzmüller, C.E., Rabe, F., Sutcliffe, G.: THF0 – The Core of the TPTP Language for Higher-Order Logic. In: Armando, A., Baumgartner, P., Dowek, G. (eds.) IJCAR 2008. LNCS (LNAI), vol. 5195, pp. 491–506. Springer, Heidelberg (2008)

5. Bradley, G.H.: Algorithms for Hermite and Smith Normal Matrices and Linear Diophantine Equations. Mathematics of Computation 25(116), 897–907 (1971)
6. Codescu, M., Horozal, F., Kohlhase, M., Mossakowski, T., Rabe, F., Sojakova, K.: Towards Logical Frameworks in the Heterogeneous Tool Set Hets. In: Mossakowski, T., Kreowski, H.-J. (eds.) WADT 2010. LNCS, vol. 7137, pp. 139–159. Springer, Heidelberg (2012)
7. Cohen, H.: A Course in Computational Algebraic Number Theory. Springer (1995)
8. Cormen, T.H., et al.: Introduction to Algorithms. McGraw-Hill (2003)
9. Cruanes, S., Hamon, G., Owre, S., Shankar, N.: Tool integration with the evidential tool bus. In: Giacobazzi, R., Berdine, J., Mastroeni, I. (eds.) VMCAI 2013. LNCS, vol. 7737, pp. 275–294. Springer, Heidelberg (2013)
10. Denney, E.: A Prototype Proof Translator from HOL to Coq. In: Aagaard, M.D., Harrison, J. (eds.) TPHOLs 2000. LNCS, vol. 1869, pp. 108–125. Springer, Heidelberg (2000)
11. Gonthier, G., Mahboubi, A.: An introduction to Small Scale Reflection in Coq. Journal of Formalized Reasoning 3(2), 95–152 (2010)
12. Gordon, M.J.C., et al.: The Right Tools for the Job: Correctness of Cone of Influence Reduction Proved Using ACL2 and HOL4. Journal of Automated Reasoning 47(1), 1–16 (2011)
13. Hendrix, J.: Matrices in ACL2. In: ACL2 2003 (2003)
14. Heras, J., Mata, G., Romero, A., Rubio, J., Sáenz, R.: Verifying a plaftorm for digital imaging: A multi-tool strategy. In: Carette, J., Aspinall, D., Lange, C., Sojka, P., Windsteiger, W. (eds.) CICM 2013. LNCS, vol. 7961, pp. 66–81. Springer, Heidelberg (2013)
15. Jacquel, M., Berkani, K., Delahaye, D., Dubois, C.: Verifying B Proof Rules Using Deep Embedding and Automated Theorem Proving. In: Barthe, G., Pardo, A., Schneider, G. (eds.) SEFM 2011. LNCS, vol. 7041, pp. 253–268. Springer, Heidelberg (2011)
16. Kaufmann, M., et al.: Computer-Aided Reasoning: An Approach. Kluwer Academic Publishers (2000)
17. Keller, C., Werner, B.: Importing HOL Light into Coq. In: Kaufmann, M., Paulson, L.C. (eds.) ITP 2010. LNCS, vol. 6172, pp. 307–322. Springer, Heidelberg (2010)
18. Naumov, P., Stehr, M.-O., Meseguer, J.: The HOL/NuPRL Proof Translator (A Practical Approach to Formal Interoperability). In: Boulton, R.J., Jackson, P.B. (eds.) TPHOLs 2001. LNCS, vol. 2152, pp. 329–345. Springer, Heidelberg (2001)
19. Nipkow, T., Paulson, L.C., Wenzel, M. (eds.): Isabelle/HOL. LNCS, vol. 2283. Springer, Heidelberg (2002)
20. Obua, S., Nipkow, T.: Flyspeck II: the basic linear programs. Annals of Mathematics and Artificial Intelligence 56(3-4), 245–272 (2009)
21. Obua, S., Skalberg, S.: Importing HOL into isabelle/HOL. In: Furbach, U., Shankar, N. (eds.) IJCAR 2006. LNCS (LNAI), vol. 4130, pp. 298–302. Springer, Heidelberg (2006)
22. Sexton, A.P., et al.: Computing with Abstract Matrix Structures. In: ISSAC 2009, pp. 325–332. ACM (2009)
23. Siekmann, J.H., Brezhnev, V., Cheikhrouhou, L., Fiedler, A., Horacek, H., Kohlhase, M., Meier, A., Melis, E., Moschner, M., Normann, I., Pollet, M., Sorge, V., Ullrich, C., Wirth, C.-P.: Proof Development with ΩMEGA. In: Voronkov, A. (ed.) CADE-18. LNCS (LNAI), vol. 2392, pp. 144–149. Springer, Heidelberg (2002)
24. Steele, G.L.: Common Lisp the Language. Digital Press (1990)
25. W3C. XSLT 2.0, http://www.w3.org/TR/xslt-xquery-serialization/

A Direct Propagation Method
in Singly Connected Causal Belief Networks
with Conditional Distributions for all Causes

Oumaima Boussarsar, Imen Boukhris, and Zied Elouedi

LARODEC, Université de Tunis, Institut Supérieur de Gestion de Tunis, Tunisia
oumaima.boussarsar@hotmail.fr, imen.boukhris@hotmail.com,
zied.elouedi@gmx.fr

Abstract. Existing algorithms of propagation in belief networks deal
with inference of observations when conditional distributions are initially
defined per edge. The aim of this paper is to propose a direct method
of causal inference of both observations and interventions on the causal
belief networks quantified with the belief function theory where conditional beliefs are defined for all parents without having to transform the
network into a junction tree. We explain how it is still possible to use
the disjunctive rule of combination DRC and the generalized Bayesian
theorem GBT to perform this propagation.

Keywords: Belief function theory, propagation, interventions, causal
belief networks.

1 Introduction

Causality plays an important role in many fields, from physics to medicine to
artificial intelligence. Indeed, causal knowledge simplifies decision-making. Interventions [6] are very useful for identifying causal relations. These latter are ex-
terior manipulations that force target variables to have specific values. However,
an observation is seeing and monitoring phenomena happening by themselves
without any manipulation on the system.

The belief function theory is adequate to formalize imperfect causal knowledge
that agents usually possess especially cases of ignorance. Accordingly, a graphical
structure allows to simply represent and reason from such causal knowledge.
Causal belief networks [3] are compact and flexible graphical representations
where arcs are interpreted as causal links. On these networks, we can compute
the effects of observations and also those of external actions.

In existing algorithms of propagation in belief networks either they are associ-
ational networks [1,11], or causal networks [4], the uncertainty is not modeled by
a conditional mass function between a node and all its parents as for Bayesian
networks, but as a set of local conditional distributions for a node and each of
its parents. In the case where the expert gives conditional distributions defined
for all parents, we must necessarily transform the network that is already simply

G.A. Aranda-Corral et al. (Eds.): AISC 2014, LNAI 8884, pp. 64–75, 2014.
© Springer International Publishing Switzerland 2014

connected (i.e., there are no two nodes that can be connected by more than one path) into a joint tree while the latter is usually used to transform multiply connected networks (i.e., arbitrary network structures) to a tree structure. Nodes in this tree are sets of variables called clusters. The propagation algorithm based on junction trees is expensive. Indeed, it depends on clusters' size.

In this paper, we propose a causal propagation method that performs directly on the initial causal belief network in the case where the conditional distributions are defined for all parents allowing to compute the effects of observations and also those of external actions. Our proposed algorithms are based on the two rules proposed by Smets [10] namely the disjunctive rule of combination (DRC) and the generalized Bayesian theorem (GBT). Moreover, we explain how these operators can be used on a set of variables.

The rest of the paper is organized as follows: in Section 2, we provide a brief background on the belief function theory. In Section 3, we recall causal belief networks. In Section 4, we explain how it is still possible to use the DRC and the GBT to perform this propagation. Inference in the presence of observations and interventions using mutilated and augmented graphs where conditional distributions are defined for all parents is described in Section 5. Section 6 concludes the paper.

2   Belief Function Theory

2.1   Definition

The theory of belief functions [8] is useful for representing uncertain knowledge. Let Θ be a finite non empty set including all the elementary events related to a given problem. These events are assumed to be exhaustive and mutually exclusive. Such Θ is called the frame of discernment. Beliefs are expressed on subsets belonging to the powerset of Θ denoted 2^{Θ}. The basic belief assignment (bba), denoted by m^{Θ} or m, is a mapping from 2^{Θ} to [0,1] such that: $\sum_{A \subseteq \Theta} m(A) = 1$. For each subset A of Θ, $m(A)$ is called the basic belief mass (bbm). It represents the part of belief exactly committed to the event A of Θ. Subsets of Θ such that $m(A) > 0$ are called focal elements. A bba is said to be certain if the whole mass is allocated to a unique singleton of Θ and Bayesian when all focal elements are singletons. If the bba has Θ as unique focal element, it is called vacuous and it represents the case of total ignorance.

The plausibility function pl quantified the maximum amount of belief that could be given to a subset A of Θ. It computes the total of masses compatible with A.

$$pl : 2^{\Theta} \rightarrow [0, 1] \quad \text{such that:}$$

$$pl(A) = \sum_{A \cap C \neq \emptyset} m(C) \tag{1}$$

The basic belief assignment can be recovered from the plausibility function as follows:

$$m(A) = \sum_{C \subseteq A} (-1)^{|A-C+1|} pl(\bar{C}) \tag{2}$$

2.2   Basic Operations

Two *bbas* m_1 and m_2 provided by two distinct and independent sources, may be aggregated using Dempster's rule of combination, denoted by $\oplus$, as follows:

$$m_1 \oplus m_2(A) = K \cdot \sum_{B \cap C = A} m_1(B)m_2(C), \forall B, C \subseteq \Theta \tag{3}$$

where $K^{-1} = 1 - \sum_{B \cap C = \emptyset} m_1(B)m_2(C)$.

Smets [9] qualified Dempster's rule of conditioning as one of the natural ingredients and the center of the transferable belief model. Upon the arrival of a new information B, the initial knowledge encoded with a mass value, $m(A)$, is revised using Dempster's rule of conditioning. $m(A|B)$ denotes the degree of belief of A in the context where B holds. It is defined as:

$$m(A|B) = \begin{cases} K. \sum_{C \subseteq \bar{B}} m(A \cup C) & \text{if } A \subseteq B, A \neq \emptyset \\ 0 & \text{if } A \not\subseteq B \end{cases} \tag{4}$$

where $K^{-1} = 1 - m(\emptyset)$.

2.3   Multi-variable Operations

When we model aspects of the real world, the *bbas* induced from experts are defined on different frames of discernment. We recall in what follows useful multi-variables operations. Let us consider in what follows, a first frame Θ and a second frame Ω. A vacuous extension is changing the referential by adding new variables. Thus, a mass function m^Θ defined on Θ will be extended to $\Theta \times \Omega$ as follows:

$$m^{\Theta \uparrow \Theta \Omega}(C) = \begin{cases} m^\Theta(A) & \text{if } C = A \times \Omega \\ 0 & \text{otherwise} \end{cases} \tag{5}$$

Given a mass distribution defined on the product space $\Theta \times \Omega$, marginalization corresponds to mapping over a subset of the product space by dropping the extra coordinates. The new belief defined on Θ, $m^{\Theta \Omega \downarrow \Theta}$ is obtained by:

$$m^{\Theta \Omega \downarrow \Theta} = \sum_{C \subseteq \Theta \times \Omega, C^{\downarrow \Theta} = A} m^{\Theta \Omega}(C), A \subseteq \Theta \tag{6}$$

Smets [10] has generalized the Bayesian theorem within the transferable belief model framework known as the Generalized Bayesian Theorem (GBT). Let us consider $pl^\Omega(c|a_i)$ and $a_i \in a$ where $a \subseteq \Theta$ and $c \subseteq \Omega$. The a posteriori plausibility distribution $pl^\Theta(a|c)$ is defined as follows:

$$pl^\Theta(a|c) = 1 - \prod_{a_i \in a} (1 - pl^C(c|a_i)) \tag{7}$$

The function that is the dual of GBT is the disjunctive rule of combination (DRC). Let us consider $pl^\Omega(c|a_i)$ and $a_i \in a$ where $a \subseteq \Theta$ and $c \subseteq \Omega$. The plausibility distribution $pl(c|a)$ is defined as follows:

$$pl^\Theta(c|a) = pl^\Theta(a|c) = 1 - \prod_{a_i \in a} (1 - pl^C(c|a_i)) \tag{8}$$

3   Causal Belief Networks

Belief networks [1,3,11] are simple and efficient tools to compactly represent uncertainty distributions. Causal reasoning can be intuitively and formally described with graphs [2,3,6]. On these networks, it is possible to predict the effects of both observations and external actions on the system. Causal belief networks [3] are used to formalize the imperfect causal knowledge. They represent an alternative to causal Bayesian networks, that allow to formalize conditional beliefs in a flexible way. It is defined on two levels:

- Qualitative level: represented by a directed acyclic graph (DAG) named G where $G = (V,E)$ in which the nodes V represent variables and edges E encode the cause-effect relations among variables. The set of parents of A is denoted by $Pa(A)$. The set of children of A is denoted by $Ch(A)$. A root is a node with no parents $(Pa(A) \neq \emptyset)$. A leaf is a node with no children $(Ch(A) \neq \emptyset)$. We will denote by R the set of roots and by L the set of leaves.
- Quantitative level: is the set of normalized *bbas* associated to each node in the graph. Conditional distributions can be defined for each variable A denoted on Θ_A in the context of its parents (either one or more than one node): $\sum_{sub_{ik} \subseteq \Theta_A} m^A(sub_{ik}|Pa(A)) = 1$

An intervention is an external action which changes some value(s) in the system and consequently will lead to different results than those found with observational data. These effects should be adequately predicted. While conditioning is used to compute the effect of observations, the "do" operator [6] is used to compute the impact of external action. Handling interventions and computing their effects on the system can be done by making changes on the structure of the belief causal network. The two equivalent methods developed were namely, belief graph mutilation method where all the edges directed to the target node will be deleted and belief graph augmentation method which consists of adding, for the target variable, a new parent variable denoted DO. Thus, the parents set of the variable A denoted PA is transformed to $Pa' = Pa \cup \{DO\}$. The DO node takes values in $do(a_i)$, $x \in \{\Theta_A \cup \{nothing\}\}$. do(nothing) represents the state of the system when no interventions are made. $do(a_i)$ means that the variable A is forced to take the certain value a_i.

4   DRC and GBT for Inference in Causal Belief Networks

4.1   Definitions

To reduce the cost of storage, the DRC and the GBT are used when the plausibility distributions are conditionally defined for singletons $(a_i \in a)$ where $a \subseteq \Theta$. The DRC is used for backward propagation. Let m^Θ be the *bba* of the parent node A which is sent to its child C.

$$pl^\Omega(c) = \sum_{a \subseteq \Theta} m^\Theta(a)(1 - \prod_{a_i \in \Theta} (1 - pl^\Omega(c|a_i))) \qquad (9)$$

The GBT can be used for forward propagation. Let m^Ω be the *bba* of the child node which is sent to its parent node A using the GBT.

$$pl^\Theta(a) = \sum_{c \subseteq \Omega} m^C(c)(1 - \prod_{a_i \in \Theta}(1 - pl^\Omega(c|a_i))) \qquad (10)$$

Exemple 1. *Let $\Theta = \{a_1, a_2\}$ and $\Omega = \{c_1, c_2\}$. m^Θ and $m^\Omega(.|a_i)$ are the a priori mass distribution. pl^Ω is computed using the DRC (see Table 1). Thanks to the mobius transformation, we can convert the plausibility distribution pl^Ω to a mass distribution m^Ω. pl^Θ is computed using the GBT using m^Ω and $m^\Omega(.|a_i)$ (see Table 2).*

<table>
<tr><td colspan="5">**Table 1.** DRC</td></tr>
<tr><td></td><td>$\{a_1\}$</td><td>$\{a_2\}$</td><td>Θ</td><td>$pl^\Omega(c)$</td></tr>
<tr><td>∅</td><td>0</td><td>0</td><td>0</td><td>**0**</td></tr>
<tr><td>$\{c_1\}$</td><td>0.5</td><td>0.2</td><td>0.6</td><td>**0.42**</td></tr>
<tr><td>$\{c_1\}$</td><td>0.5</td><td>0.8</td><td>0.9</td><td>**0.78**</td></tr>
<tr><td>Ω</td><td>1</td><td>1</td><td>1</td><td>**1**</td></tr>
<tr><td>m^Θ</td><td>0.2</td><td>0.4</td><td>0.4</td><td></td></tr>
</table>

<table>
<tr><td colspan="5">**Table 2.** GBT</td></tr>
<tr><td></td><td>m^Ω</td><td>$\{a_1\}$</td><td>$\{a_2\}$</td><td>Θ</td></tr>
<tr><td>∅</td><td>0</td><td>0</td><td>0</td><td>0</td></tr>
<tr><td>$\{c_1\}$</td><td>0.2</td><td>0.5</td><td>0.2</td><td>0.6</td></tr>
<tr><td>$\{c_1\}$</td><td>0</td><td>0.5</td><td>0.8</td><td>0.9</td></tr>
<tr><td>Ω</td><td>0.8</td><td>1</td><td>1</td><td>1</td></tr>
<tr><td>$pl^\Theta(a)$</td><td></td><td>**0.9**</td><td>**0.84**</td><td>**0.92**</td></tr>
</table>

4.2 Propagating Distributions for All Parents

In this section, we explain that is possible to use the GBT and the DRC for propagation when the relations between nodes are not binary (i.e., conditional distributions are defined for all parents).

Given a set of variables $(A_1, A_2, \ldots, A_i)$ which are parent nodes of a variable C. To apply the GBT and the DRC, we will consider the m-tuple of the cartesian product of the parent nodes. Accordingly, the first component of the i-tuple belongs to A_1, the second A_2 and the i-th to A_i.

Each i-tuple will be considered as a singleton. To reduce the cost of storage and facilitate to experts to express their beliefs, the plausibility distributions of C will be defined and stored in the context of singletons of the cartesian product $(A_1, A_2, \ldots, A_i)$.

Example 2. *Let us consider the following directed causal belief network in Figure 1 where A and B are the parents of C. For the sake of simplicity, all the variables used in this example are binary. The DRC and the GBT can be applied for singletons $(a_i \in a)$. Since the conditional mass distributions of C are defined for all parents A and B, conditional mass distributions will be defined for subsets. So, we have to use the cartesian product of the parents node A and B ($A \times B = \{a_1b_1, a_1b_2, a_2b_1, a_2b_2\}$). When applying the DRC and the GBT, the conditional plausibility distributions of C are saving according to singletons of the cartesian product $A \times B$.*

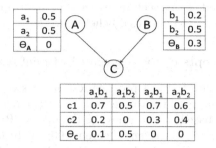

	a_1b_1	a_1b_2	a_2b_1	a_2b_2
c1	0.7	0.5	0.7	0.6
c2	0.2	0	0.3	0.4
Θ_c	0.1	0.5	0	0

Fig. 1. Belief network where distributions are defined for all parents

5   Inference in Singly Connected Causal Belief Networks

The impact of a new piece of information on the remaining variables can be found by first computing the joint distribution and then making marginalization by dropping the extra coordinates. This method is not suitable when the number of variables becomes substantial. To solve this problem, equivalent local computations have been proposed [5,7].

Existing algorithms only deal with the propagation of observational data in belief networks [1,12] where distributions are defined per single parent. To ensure propagation in the case where distributions are defined for all parents, we have to transform the initial network into a junction tree even if this technique is usually used to transform a multiply network into a tree structure. This transformation is expensive and the propagation algorithm depends on clusters' size. To tackle these problems, we propose a direct method of propagation in causal belief networks where distributions are defined for all parents. The proposed method consists of updating the belief mass of each node. If the node has more than one parent, we need to combine the distributions of parent nodes using the vacuous extension to the product space of variables representing the parent nodes. In our approach, the combined distribution is stored in a fictional node allowing message passing to its child node.

Causal propagation consists of finding the influence of an intervention or an observation on the remaining variables of the system. This is done through message passing between variables. When receiving a message each node X updates both local vectors; the vector $\pi(x_1, ..., x_n)$ concerning messages received by its parents and the vector $\lambda(x_1, ..., x_n)$ concerning messages received by its children. Each node sends and receives messages from each of its neighbors. The local message-passing between variables is based on two kinds of messages. The π-message is a message sent from a parent node to a child node and the λ-message is a message sent from a child node to a parent node.

In this section, we will first introduce the basic concepts of propagation in belief networks. Then, we explain how to compute the mass distribution of the fictional node. At the end of this section, we present algorithms for propagating observations and interventions in causal belief networks.

5.1   The Basic Concepts of Propagation in Belief Networks

Message passing is termed forward propagation or backward propagation depending on the direction in which the message is circulated. The causal belief inference algorithm are based on two rules the Disjunctive Rule of Combination (DRC) and the Generalized Bayesian Theorem (GBT). In fact, the algorithm consists of two phases: propagation down (backward) and propagation up (forward).

Let us consider the two nodes A and C where A is the parent of C. The message sent from A to C is a π-message computed using the DRC and the message sent from C to A is a λ-message computed using the GBT.

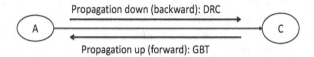

Fig. 2. Propagation process between A and C

5.2   Computation of the Mass Distribution of the Fictional Node

To perform the propagation in a simply connected network where nodes are originally defined in the context of all parents, we will combine the mass distributions of parent nodes. These distributions may correspond to the a priori distributions in the case of root nodes or posteriori distributions to the other nodes computed using the GBT and the DRC. The result of this combination will be stored in a table associated with a fictional node. The mass distribution of the fictional node is computed using the Dempster rule of combination after the extension of the mass distributions of different parents Pa(A) of the visited node A to a joint space using the vacuous extension. The mass distribution of the fictional node is denoted as $m_{fictional}$. It is computed as follows:

$$m_{fictional} = \oplus_{A \in Pa(A)}(m^{A \uparrow Pa(A)}) \tag{11}$$

Once we combined the distribution of the node parents, we can make the propagation up and down using the two operators DRC and GBT.

Example 2 (Continued). *Let us continue with the same network presented in Figure 1. Let m^A and m^B be the mass distributions of the two nodes A and B.*

To combine these nodes into a fictional node AB, we have to use the vacuous extension to extend A and B to a joint space $A \times B$ (see Table 3 and Table 4). The mass distribution of the fictional node is then computed using the Dempster rule of combination (see Figure 3).

Table 3. $m^{A \uparrow AB}$

$a_1 \times \Theta_B$	0.5
$a_2 \times \Theta_B$	0.5
$\Theta_A \times \Theta_B$	0

Table 4. $m^{B \uparrow AB}$

$b_1 \times \Theta_A$	0.2
$b_2 \times \Theta_A$	0.5
$\Theta_B \times \Theta_A$	0.3

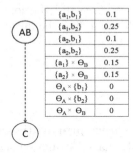

$\{a_1, b_1\}$	0.1
$\{a_1, b_2\}$	0.25
$\{a_2, b_1\}$	0.1
$\{a_2, b_2\}$	0.25
$\{a_1\} \times \Theta_B$	0.15
$\{a_2\} \times \Theta_B$	0.15
$\Theta_A \times \{b_1\}$	0
$\Theta_A \times \{b_2\}$	0
$\Theta_A \times \Theta_B$	0

Fig. 3. A causal belief network with a fictional node

5.3 Propagation of Observations in Causal Belief Networks

We propose in this section a direct propagation algorithm in singly connected causal belief networks where conditional beliefs are defined for all parents. The causal direct propagation of observations consists of two steps: the propagation down and the propagation up. A post-order (in direction of leaves) and a pre-order (in direction of roots) will be defined to propagate information backward and forward respectively.

Algorithm. Propagation down

For each A ∈ post-order
   If A ∉ R
      Combine the masses of its parents using the vacuous extension.
      Store the combined distribution in a fictional node.
      Pass a message π from the fictional node to A using the DRC.
      Compute its mass distribution.
      Send a message to its child C.
      Marginalization: find the initial mass of the parent nodes.
   End if
End for

Algorithm. Propagation up

For each A ∈ pre-order
  If A ∉ R
    Combine the masses of parents.
    Store the combined distribution in a fictional node.
    Send a message λ to the fictional node using the GBT.
    Compute its mass distribution.
    Marginalization: find the initial mass of the parent nodes.
  End if
End for

Algorithm. Direct propagation of observations
  Updating the mass distribution of the node concerned by the observation.
  Propagation down.
  Propagation up.

Each node A computes its mass distribution by combining the two values π and λ using this formula:

$$m \leftarrow \pi_A \oplus \lambda_A \tag{12}$$

Example 2 (Continued). *Let us continue with same example. Propagation consists of sending a message π from the mass distribution of the fictional node that is resulting from the combination of distributions of A and B to the node C. This latter computes its message π using the DRC (Equation 9). The new distribution of node C is as follows: $c_1 = 0.73$, $c_2 = 0.17$, $\Theta_C = 0.1$*
Then, the node C sends a message λ to the fictional node AB which in turn computes the new value λ using the GBT (Equation 10). The results are subsets of the cartesian product of $A \times B$. The distribution of AB as follows:
$m^{AB}(\{(a_1, b_1)\}) = 0.0152$, $m^{AB}(\{(a_1, b_2)\}) = 0.0185$, $m^{AB}(\{(a_2, b_1)\}) = 0.026$, $m^{AB}(\{(a_2, b_2)\}) = 0.0404$, $m^{AB}(\{(a_1, b_1), (a_1, b_2)\}) = 0.0744$, $m^{AB}(\{(a_1, b_1), (a_2, b_1)\}) = 0.0065$, $m^{AB}(\{(a_1, b_1), (a_2, b_2)\}) = 0.01$, $m^{AB}(\{(a_1, b_2), (a_2, b_1)\}) = 0.0434$, $m^{AB}(\{(a_1, b_2), (a_2, b_2)\}) = 0.0279$, $m^{AB}(\{(a_2, b_1), (a_2, b_2)\}) = 0.0173$, $m^{AB}(\{(a_1, b_1), (a_1, b_2), (a_2, b_1)\}) = 0.1734$, $m^{AB}(\{(a_1, b_1), (a_2, b_1), (a_2, b_2)\}) = 0.0043$, $m^{AB}(\{(a_1, b_1); (a_1, b_2), (a_2, b_2)\}) = 0.1115$, $m^{AB}(\{(a_1, b_2), (a_2, b_1); (a_2, b_2)\}) = 0.065$, $m^{AB}(\{(a_1, b_1), (a_1, b_2), (a_2, b_1), (a_2, b_2)\}) = 0.3662$

After computing the distributions of the fictional node AB, it is possible to compute the mass distribution of A and B by applying the marginalization $m^{AB \downarrow A}$ where $m^{AB \downarrow A}(a_1) = 0.1081$, $m^{AB \downarrow A}(a_2) = 0.0837$ and $m^{AB \downarrow A}(\Theta_A) = 0.8082$ and $m^{AB \downarrow B}$ where $m^{AB \downarrow B}(b_1) = 0.0477$, $m^{AB \downarrow B}(b_2) = 0.0868$ and $m^{AB \downarrow B}(\Theta_B) = 0.8655$ since the propagation is ensured between a node and its neighbors.

5.4 Propagation of Interventions in Causal Belief Networks

Handling interventions can be done using graph augmentation and graph mutilation methods.

- **Propagation in the mutilated graph**

  Propagation in this graph consists of two steps: the mutilation step where the distribution of concerned by the intervention becomes a certain one (see Figure 4) and the propagation step using the direct causal propagation algorithm presented in Section 5.3.

  > **Algorithm. Direct propagation using the mutilated based approach**
  > Cutting all edges pointing to the node concerned by the intervention C.
  > C computes its new marginal which becomes a certain bba.
  > Propagation down.
  > Propagation up.

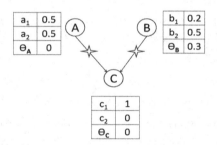

Fig. 4. A causal belief mutilated graph

- **Propagation in the augmented graph**

  Since adding the "DO" node, the conditional distribution of the node concerned by the intervention A given all parents must be updated. Hence, the graph augmentation method allows to represent the effect of observations when the DO node is taking the value nothing . When the DO node is taking the $do(a_i)$, we make a certain action which succeeds to put its target at a precise value by making it completely independent of its original causes. Thus, the distribution of A is a certain bba. Let $Pa(A)$ be the parents of the A except the DO node, the conditional distribution of the A is defined as follows:

  $$m(a_k|Pa(A), do(x)) = \begin{cases} 1 & \text{if } x = a_i \\ 0 & \text{if } x \neq a_i \\ m(a_k|Pa(A), do(x)) & x = nothing \end{cases} \tag{13}$$

Propagation in this graph consists of two steps: the augmentation step where the conditional distribution of the node concerned by the intervention becomes a certain due to addition of the node DO and the propagation step using the using the direct causal propagation algorithm presented in Section 5.3.

Algorithm. Direct propagation using the augmented based approach
  Add the node DO as a parent of the node concerned by the intervention C.
  Updating the conditional mass distribution of C using Equation 13.
  Propagation down.
  Propagation up.

Example 3. *Let us consider the network presented in Figure 5 which illustrates a causal belief augmented graph on which an intervention $do(c_1)$ forces the variable C to take the specific value c_1. The conditional bba of C given its parents DO and A are defined using the Equation 13.*

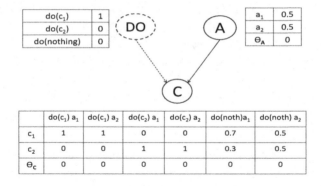

Fig. 5. A causal belief augmented graph

In the case where the intervention $do(c_1)$ forces the variable C to take the specific value c_1, the mass distribution of the node C after the propagation process using the DRC is a certain bba. When the variable takes the value nothing, the bba is the same where there is no intervention. The parent nodes are extended to the joint space $DO \times A$ using the vacuous extension and then combined using the Dempster rule of combination. The results of the propagation are as follows: $do(c_1)$: $c_1=1$, $c_2=0$, $\Theta_C=0$; $do(c_2)$: $c_1=0$, $c_2=1$, $\Theta_C=0$; $do(nothing)$: $c_1=0.6$, $c_2=0.4$, $\Theta_C=0$

6   Conclusion

In this paper, we explained how we can still use the DRC and the GBT rules for propagation in causal belief networks even if the distributions are defined for all parents. We proposed a method acting directly on the network without having to go through the transformation into a junction tree. The proposed algorithms deal with the inference of observations and interventions in the augmented and mutilated graphs. As future work, we intend to treat inference in multiply connected causal belief networks. Inference in causal belief networks can be used in several applications like those allowing the intrusion detection and or ensuring system reliability.

References

1. Ben Yaghlane, B., Mellouli, K.: Inference in directed evidential networks based on the transferable belief model. International Journal Of Approximate Reasoning 48, 399–418 (2008)
2. Benferhat, S., Smaoui, S.: Possibilistic causal networks for handling interventions: A new propagation algorithm. In: AAAI Conference on Artificial Intelligence (AAAI), pp. 373–378. AAAI Press (2007)
3. Boukhris, I., Elouedi, Z., Benferhat, S.: Dealing with external actions in causal belief networks. International Journal Of Approximate Reasoning, 978–999 (2013)
4. Boussarsar, O., Boukhris, I., Elouedi, Z.: Representing interventional knowledge in causal belief networks: Uncertain conditional distributions per cause. In: Laurent, A., Strauss, O., Bouchon-Meunier, B., Yager, R.R. (eds.) IPMU 2014, Part III. CCIS, vol. 444, pp. 223–232. Springer, Heidelberg (2014)
5. Pearl, J.: Probabilistic Reasoning in Intelligent Systems: Networks of Plausible Inference. Morgan Kaufmann Pub., San Mateo (1988)
6. Pearl, J.: Causality: Models, Reasonning and Inference. Cambridge University Press (2000)
7. Shachter, R.D.: Probabilistic inference and influence diagrams. Operations Research 36, 589–604 (1988)
8. Shafer, G.: A Mathematical Theory of Evidence. Princeton Univ. Press, Princeton (1976)
9. Smets, P.: The combination of evidence in the transferable belief model. IEEE Pattern Analysis and Machine Intelligence 12, 447–458 (1990)
10. Smets, P.: Jeffrey's rule of conditioning generalized to belief functions. In: Uncertainty in Artificial Intelligence, pp. 500–505 (1993)
11. Xu, H., Smets, P.: Evidential reasoning with conditional belief functions. In: Uncertainty in Artificial Intelligence, pp. 598–606 (1994)
12. Xu, H., Smets, P.: Reasoning in evidential networks with conditional belief functions. International Journal of Approximate Reasoning 14, 155–185 (1996)

From Declarative Set Constraint Models
to "Good" SAT Instances

Frédéric Lardeux[1] and Eric Monfroy[2]

[1] Université d'Angers, France
Frederic.Lardeux@univ-angers.fr
[2] LINA, UMR CNRS 6241, TASC INRIA, Université de Nantes, France
Eric.Monfroy@univ-nantes.fr

Abstract. On the one hand, Constraint Satisfaction Problems allow one to declaratively model problems. On the other hand, propositional satisfiability problem (SAT) solvers can handle huge SAT instances. We thus present a technique to declaratively model set constraint problems, to reduce them, and to encode them into "good" SAT instances. We illustrate our technique on the well-known nqueens problem. Our technique is simpler, more expressive, and less error-prone than direct hand modeling. The SAT instances that we automatically generate are rather small w.r.t. hand-written instances.

1 Introduction

Most of combinatorial problems can be formulated as Constraint Satisfaction Problems (CSP) [18]. A CSP is defined by some variables and constraints between these variables. Solving a CSP consists in finding assignments of the variables that satisfy the constraints. One of the main strength of CSP is declarativity and expressiveness: variables can be of various types (finite domains, floating point numbers, sets, ...) and constraints as well (linear arithmetic constraints, set constraints, non linear constraints, Boolean constraints, ...). Moreover, the so-called global constraints not only improve solving efficiency but also expressiveness: they propose new constructs and relations such as *alldifferent* (to enforce that all the variables of a list have different values), *cardinality* (to link a set to its size), ...

On the other hand, the propositional satisfiability problem (SAT) [8] is restricted (in terms of expressiveness) to Boolean variables and propositional formulae. Coding set constraints directly into SAT is a tedious tasks (see for example [19] or [9]). Moreover, when one wants to optimize its model in terms of variables and clauses this quickly leads to very complicated and unreadable models in which errors can easily appear. However, SAT solvers can now handle huge SAT instances (millions of variables). It is thus attractive to 1) encode CSPs into SAT (e.g., [3,5]) in order to benefit from the declarativity and expressiveness of CSP and the power of SAT, and 2) introduce more declarativity into SAT, e.g., with global constraints such as alldifferent [12], or cardinality [4].

G.A. Aranda-Corral et al. (Eds.): AISC 2014, LNAI 8884, pp. 76–87, 2014.

Various systems of set constraints (either specialized systems [13], libraries for constraint programming systems such as [10], or the set constraint library of CHOCO [1]) have been designed and it has been shown that numerous problems can easily be modeled with set constraints.

In this paper we are concerned with the transformation of set constraints into SAT instances: we often refer to this transformation as "encoding". In [11], we presented encoding rules that are directly applied on the CSP set constraints. However, we have noticed that some supports of sets (i.e., elements that are possibly in this set) could be reduced (some elements can be removed from the supports without loosing any solution). Thus, the generated SAT instances include non necessary information.

It is inconceivable to force the user to write reduced CSP models: first, because it is a tedious and error-prone task; and second, it may be impossible to see all the relations between the sets (more especially when working on sets which supports are not yet declared). Thus, our approach consists in providing:

- a simple but complete, declarative, and expressive set language for easily modeling problems with constraints such as intersection, union, cardinal of sets, ...
- a set of reduction rules ($\Rightarrow_{red}$) to reduce CSP models. In fact, these rules define constraint propagation [2] for sets and elements and make the model Generalized Arc Consistent [14].
- a set of encoding rules ($\Leftrightarrow_{enc}$) that convert CSP constraints into SAT instances.

In this paper, we illustrate our approach with the famous nqueen problem. Moreover, we have tried our technique on various problems (e.g., Social Golfer Problem [11], Sudoku, Car-sequencing) and the SAT instances which are automatically generated have a complexity similar to the complexity of improved hand-written SAT formulations, and their solving with a SAT solver (in our case Minisat [6]) is efficient compared to other SAT approaches.

Compared to [11], the $\Rightarrow_{red}$ reduction rules enable us:

- to simplify the encoding rules ($\Leftrightarrow_{enc}$): indeed, some transformation cases become unuseful; the encoding rules become even more simple and readable;
- to obtain even smaller SAT instances, in terms of clauses and variables; these problems are solved faster;
- to tackle and solve larger problems that we were unable to encode using only our previous encoding rules (for size reasons).

We can compare our work with SAT encoding techniques such as [3] and [5]. These works make a relation between CSP solving and SAT solving in terms of properties such as consistencies for finite domain variables and constraints. In this article, we are concerned with a different type of constraints (i.e., set constraints) and we try to obtain small SAT instances that are also well-suited for standard SAT solvers. Our approach is similar to [12] in which alldifferent global constraints and overlapping alldifferent constraints are handled declaratively before being encoded automatically in SAT using rewrite rules. Note also that we use the work of [4] about the *cardinality* global constraint in order to perform the encoding of set cardinality. Our goal is not to compete with standard set solvers, but to introduce set constraints into SAT.

In the next section (Section 2) we present the CSP set constraint language. In Section 3 we show the rules to reduce models. Section 4 presents our new rule-based system for encoding set constraints into SAT. Section 5 illustrates our approach on the nqueen problem. We finally conclude in Section 6.

2   Set Constraint

2.1   Universe and Supports

In order to encode set constraints into SAT, we consider 3 notions: *universe*, *support*, and *domain*. Informally, the universe is the set of all elements that are considered in a model of a given problem; the support $\mathcal{F}$ of a set F appearing in this model is a set of possible elements of F (i.e., $\mathcal{F}$ is a superset of F); and the domain of an element variable is a set of possible values for this element.

Definition 1. *Let P be a problem, and M be a model of P in $\mathcal{L}$, i.e., a description of P from the natural language to the language of constraints $\mathcal{L}$.*

- *The universe $\mathcal{U}$ of M is a finite set of constants.*
- *The support $\mathcal{F}$ of the set F of the model M is a subset of the universe $\mathcal{U}$. $\mathcal{F}$ represents the constants of $\mathcal{U}$ that can possibly be elements of F:*

$$F \subseteq \mathcal{F} \subseteq \mathcal{U} \quad and \quad F \in \mathcal{P}(\mathcal{F})$$

where $\mathcal{P}(\mathcal{F}) = \{A | A \subseteq \mathcal{F}\}$ is the power set of $\mathcal{F}$. We say that F is over $\mathcal{F}$.
- *The domain D_x of a variable element x is a subset of the universe $\mathcal{U}$; D_x represents the elements of $\mathcal{U}$ that are possible values (i.e., constants) for x.*

Note that each element of $\mathcal{U} \setminus \mathcal{F}$ cannot be an element of F. In the following, we denote sets by upper-case letters (e.g., F) and their supports by calligraphic upper-case letters (e.g., $\mathcal{F}$). Variable elements are represented by lower-case letters (e;g., x) and their domain by D indexed by the variable name (e.g., D_x). When there is no confusion, we shorten "the set F of the model M" to "F".

Consider a model M with a universe $\mathcal{U}$, and a set F over $\mathcal{F}$. For each element x of $\mathcal{F}$, we consider a Boolean variable $x_{\mathcal{F}}$ which is true if $x \in F$ and false otherwise. We call the set of such variables the support variables for F in $\mathcal{F}$. In the following, we write $x_{\mathcal{F}}$ for $x_{\mathcal{F}} = true$ and $\neg x_{\mathcal{F}}$ for $x_{\mathcal{F}} = false$.

Example 1. Let $\mathcal{U} = \{x, y, z, t\}$ be the universe of a model M, and $\mathcal{F} = \{x, y, t\}$ be the support of a set F of M. Then, we have 3 Boolean variables $x_{\mathcal{F}}$, $y_{\mathcal{F}}$, and $t_{\mathcal{F}}$ corresponding respectively to x, y, and t to represent F. By definition, $z \notin F$ and there is no $z_{\mathcal{F}}$ variable; and x, y, t can possibly be in F. Consider now that $F = \{x, y\}$. Then, we have $x_{\mathcal{F}}$, $y_{\mathcal{F}}$, and $\neg t_{\mathcal{F}}$.

2.2 Syntax

In order to declare objects, we use the following declarations:

- $Universe(U)$ is used to declare the universe as the set U;
- $Set(F, \mathcal{F})$: declares a set F together with its support $\mathcal{F}$.;
- $Element(x, D_x)$: creates a variable x of type element with its domain D_x.

Consider F, G, H, and F_i (i ranging from 1 to n) being sets, and x being an element. We consider the following usual (CSP) set constraints:

element (dis)equality	$x = y$	$(x \neq y)$
(non)membership	$x \in F$	$(x \notin F)$
set (dis)equality	$F = G$	$(F \neq G)$
intersection	$H = F \cap G$	
union	$H = F \cup G$	
inclusion	$F \subseteq G$	
difference	$H = F \setminus G$	
multi-intersection	$F = \bigcap_{i=1}^{n} F_i$	
multi-union	$F = \bigcup_{i=1}^{n} F_i$	
cardinality$\{=, <, >\}$	$|F| \{=, <, >\} k$	

More constraints could be defined, but they can be deduced from these basic constraints. A model for a problem is given by:

1. a universe;
2. some sets together with their supports;
3. some variable elements with their domains;
4. some constraints between sets and elements.

3 Reducing Supports

Support sizes are a crucial parameter for the sizes of generated SAT instances. Moreover, it is quite complicated (and sometimes impossible) to write a model with "reduced" supports. For example, consider 3 sets: $Set(G, \{1, 2, \ldots, 10000\})$, $Set(F, \{9999, \ldots, 20000\})$, and $Set(H, \{5000, \ldots, 25000\})$. Latter in the model, let consider that the constraint $H = F \cup G$ appear. Then, the support of H can be reduced to $\{5000, \ldots, 20000\}$, and the support of G can be reduced to $\{5000, \ldots, 10000\}$.

We thus consider some reduction rules $\Rightarrow_{red}$ to reduce domains and supports w.r.t. constraints. These rules remove elements of the supports and domains that cannot participate in any solution to the problem. We first start with failure case, i.e., cases that do not lead to any solution.

Failures. Rule 1 causes a fail when the domain of a variable is empty. Rule 2 leads to a fail when the imposed cardinality is higher than the size of the support of the set. Rule 3 is similar for inequality about cardinal.

$$D_x = \emptyset \Rightarrow_{red} fail \tag{1}$$

$$|F| = k \Rightarrow_{red} fail \text{ if } |\mathcal{F}| < k \tag{2}$$

$$|F| > k \Rightarrow_{red} fail \text{ if } |\mathcal{F}| \leq k \tag{3}$$

Domain Reduction. Rule 4 reduces the domains of two equal variables. When 2 variables are disequal, Rule 5 reduces the domain of the second variable when the domain of the first one is restricted to a singleton ($\{v_x\}$). The domain of a variable x is reduced by Rule 6 w.r.t. the support of a set F in which x must appear (constraint $x \in F$):

$$x = y \Rightarrow_{red} \begin{cases} D_x \leftarrow D_x \cap D_y, \\ D_y \leftarrow D_x \cap D_y \end{cases} \tag{4}$$

$$x \neq y, D_x = \{v_x\} \Rightarrow_{red} D_y \leftarrow D_y \setminus \{v_x\} \tag{5}$$

$$x \in F \Rightarrow_{red} D_x \leftarrow D_x \cap F \tag{6}$$

Support Reduction. When 2 sets must be equal, Rule 7 reduces their supports to their intersection. Intersection constraint enables to reduce the domain of the set intersection (Rule 8) whereas union constraint may reduce the supports of the 3 sets appearing in the constraint (Rule 9). Inclusion constraint only reduces the support of the included set (Rule 10). Difference constraint may reduce 2 supports of the 3 sets (Rule 11). Rules 12 and 13 are similar to Rules 9 and 8 for multi-union and multi-intersection constraints.

$$F = G \Rightarrow_{red} \mathcal{F} \leftarrow \mathcal{F} \cap \mathcal{G}, \quad \mathcal{G} \leftarrow \mathcal{G} \cap \mathcal{F} \tag{7}$$

$$H = F \cap G \Rightarrow_{red} \mathcal{H} \leftarrow \mathcal{H} \cap \mathcal{F} \cap \mathcal{G} \tag{8}$$

$$H = F \cup G \Rightarrow_{red} \mathcal{H} \leftarrow \mathcal{H} \cap (\mathcal{F} \cup \mathcal{G}), \quad \mathcal{F} \leftarrow \mathcal{F} \cap \mathcal{H}, \quad \mathcal{G} \leftarrow \mathcal{G} \cap \mathcal{H} \tag{9}$$

$$F \subseteq G \Rightarrow_{red} \mathcal{F} \leftarrow \mathcal{F} \cap \mathcal{G} \tag{10}$$

$$H = F \setminus G \Rightarrow_{red} \mathcal{H} \leftarrow \mathcal{H} \cap \mathcal{F}, \quad \mathcal{F} \leftarrow \mathcal{F} \cap \mathcal{H} \tag{11}$$

$$H = \bigcup_{i=1}^{n} F_i \Rightarrow_{red} \mathcal{H} \leftarrow \mathcal{H} \cap \left(\bigcup_{i=1}^{n} \mathcal{F}_i\right), \quad \forall i \in [1..n] \; \mathcal{F}_i \leftarrow \mathcal{F}_i \cap \mathcal{H}, \tag{12}$$

$$H = \bigcap_{i=1}^{n} F_i \Rightarrow_{red} \mathcal{H} \leftarrow \mathcal{H} \cap \left(\bigcap_{i=1}^{n} \mathcal{F}_i\right) \tag{13}$$

Rule Application. $\Rightarrow_{red}$ rules can be seen as filtering (or reduction) functions in constraint programming. They can thus be applied by a fixed-point algorithm such as chaotic iterations [2,16,15]: since the rules have the required properties (monotonic decreasing and idempotent), termination is ensured.

In fact, these rules define constraint propagation for sets and elements. Moreover, they enforce GAC (Generalised Arc Consistency [14]), i.e., the supports and domains cannot be reduced anymore using a single constraint without loosing solution local to this constraint. Since this is not the focus of this paper, we don't give here the proof, but just the basis: with respect to GAC, variable domains (in terms of constraint programming) are the domains of the variable elements, and the power-set of the supports for sets.

4   The $\Leftrightarrow_{enc}$ Encoding Rules

We can now define the encoding of our CSP set constraints into SAT. In the following, we consider three sets F, G, and H respectively defined on the supports $\mathcal{F}$, $\mathcal{G}$ and $\mathcal{H}$ of the universe $\mathcal{U}$, and for each $x \in \mathcal{U}$ the various Boolean variables $x_{\mathcal{F}}$, $x_{\mathcal{G}}$, and $x_{\mathcal{H}}$ as defined before. $|G|$ denotes the cardinality of the set G.

Contrary to [11], we consider here that supports and domains are reduced using $\Rightarrow_{red}$ rules. Allowing the supports to be non reduced eases the modeling process: indeed, one does not have to compute the reduced support and can use a superset of it or the universe; then, supports are reduced automatically by the $\Rightarrow_{red}$ rules and the $\Leftrightarrow_{enc}$ encoding rules can generate smaller SAT instances.

The clauses that are generated by these rules are of the form $\forall x \in \mathcal{F}, \phi(x_{\mathcal{F}})$ which denotes the $|\mathcal{F}|$ formulae $\phi(x_{\mathcal{F}})$ built for each element x of the support $\mathcal{F}$ of F (x refers to the element of the universe/support, and $x_{\mathcal{F}}$ to the variable representing x for the set F).

$Element(x, D_x)$ and $set(F, \mathcal{F})$ enable to create the required SAT variables: as many variables as the support for a set, and as many as the domain for a variable element. In the following, we present rules for set constraint encodings with: first, the set constraint, then its encoding in SAT (i.e., some clauses linking the SAT variables), and finally, the number of clauses generated.

Element Variable. This encoding rule enforces each element variable to have one and only one value from its domain:

$$Element(v, D_v) \Leftrightarrow_{enc} \forall x \in D_v, \bigvee_{x \in D_v} (\wedge_{y \in D_v, x \neq y}(\neg y_v) \wedge x_v) \quad |D_v|^2 \text{ bin. clauses}$$

Element Variable (dis) Equality. let us recall that after application of $\Rightarrow_{red}$ rules on $v = w$, v and w have the same domain. This is not the case for $v \neq w$.

$$v = w \quad \Leftrightarrow_{enc} \quad \forall x \in D_v, x_v \leftrightarrow x_w \quad 2.|D_v| \text{ binary clauses}$$

$$v \neq w \quad \Leftrightarrow_{enc} \quad \begin{cases} \forall x \in D_v, x_v \to \neg x_w & 2.|D_v| \text{ binary clauses} \\ \forall x \in D_w, x_w \to \neg x_v & 2.|D_w| \text{ binary clauses} \end{cases}$$

Membership Constraint. This constraint enforces the element v to be in the set F: if $x \in \mathcal{F}$ (x is in the support of F), then the corresponding support variable must be true (i.e., $x_{\mathcal{F}}$). The constraint $x \notin F$ can be similarly defined.

$$v \in F \Leftrightarrow_{enc} \forall x \in D_v, x_v \to x_{\mathcal{F}} \quad |D_v| \text{ binary clauses}$$

$$v \notin F \Leftrightarrow_{enc} \forall x \in D_v \cap \mathcal{F}, x_v \to \neg x_{\mathcal{F}} \wedge x_{\mathcal{F}} \to \neg x_v \quad 2.|D_v| \text{ binary clauses}$$

Set (Dis) Equality Constraint. After reduction, 2 equal sets G and F have the same support. Thus, the encoding for the equality constraint is:

$$F = G \iff_{enc} \forall x \in \mathcal{F}, \ x_{\mathcal{F}} \leftrightarrow x_{\mathcal{G}} \qquad 2.|\mathcal{F}| \text{ binary clauses}$$

The constraint $F \neq G$ is satisfied when at least one variable of the intersection of the 2 sets is different in F and G, or when a variable appearing in the support of F and not in the one of G is true (and vice-versa):

$$F \neq G \iff_{enc} \left(\bigvee_{x \in \mathcal{F} \cap \mathcal{G}} x_{\mathcal{F}} \leftrightarrow \neg x_{\mathcal{G}}\right) \vee \left(\bigvee_{x \in \mathcal{F} \setminus \mathcal{G}} x_{\mathcal{F}}\right) \vee \left(\bigvee_{x \in \mathcal{G} \setminus \mathcal{F}} x_{\mathcal{G}}\right)$$
$$2.|\mathcal{F} \cup \mathcal{G}| \text{ clauses of size } 2 + |\mathcal{F} \cap \mathcal{G}| - |\mathcal{F} \cup \mathcal{G}|$$

Intersection Constraint. Let H be the intersection of two sets G and F: the reduced support of H is included in the intersection of the supports of G and F.

- for the elements of $\mathcal{F} \cap \mathcal{G} \cap \mathcal{H}$: a support variable of H is true if and only if this variable is in F and G;
- for the elements of $(\mathcal{F} \cap \mathcal{G}) \setminus \mathcal{H}$: since such an element cannot be in H, it must not be in F or in G.

$$F \cap G = H \quad \iff_{enc}$$
$$\begin{cases} \forall x \in \mathcal{F} \cap \mathcal{G} \cap \mathcal{H}, \ x_{\mathcal{F}} \wedge x_{\mathcal{G}} \leftrightarrow x_{\mathcal{H}} & |\mathcal{F} \cap \mathcal{G} \cap \mathcal{H}| \text{ ternary clauses} \\ & +2.|\mathcal{F} \cap \mathcal{G} \cap \mathcal{H}| \text{ binary clauses} \\ \forall x \in (\mathcal{F} \cap \mathcal{G}) \setminus \mathcal{H}, \ \neg x_{\mathcal{F}} \vee \neg x_{\mathcal{G}} & |(\mathcal{F} \cap \mathcal{G}) \setminus \mathcal{H}| \text{ binary clauses} \end{cases}$$

Union Constraint. More cases must be considered for this constraints:

- for the elements of $\mathcal{F} \cap \mathcal{G} \cap \mathcal{H}$: a support variable of H is true if and only if this variable is in F or in G; this is the trivial case;
- for the elements of $(\mathcal{F} \cap \mathcal{H}) \setminus \mathcal{G}$: this case is a reduction of the previous one but it is however equivalent; since such an element x is not in the support of G then $x_{\mathcal{G}}$ does not exist, and x is in H if and only if it is in F; note that the generated clauses are exactly the same removing $x_{\mathcal{G}}$;
- for the elements of $(\mathcal{G} \cap \mathcal{H}) \setminus \mathcal{F}$: this is the symmetrical case for G;

$$F \cup G = H \quad \iff_{enc}$$
$$\begin{cases} \forall x \in \mathcal{F} \cap \mathcal{G} \cap \mathcal{H}, \ x_{\mathcal{F}} \vee x_{\mathcal{G}} \leftrightarrow x_{\mathcal{H}} & |\mathcal{F} \cap \mathcal{G} \cap \mathcal{H}| \text{ ternary clauses} \\ & +2.|\mathcal{F} \cap \mathcal{G} \cap \mathcal{H}| \text{ binary clauses} \\ \forall x \in (\mathcal{F} \cap \mathcal{H}) \setminus \mathcal{G}, \ x_{\mathcal{F}} \leftrightarrow x_{\mathcal{H}} & 2.|(\mathcal{F} \cap \mathcal{H}) \setminus \mathcal{G}| \text{ binary clauses} \\ \forall x \in (\mathcal{G} \cap \mathcal{H}) \setminus \mathcal{F}, \ x_{\mathcal{G}} \leftrightarrow x_{\mathcal{H}} & 2.|(\mathcal{G} \cap \mathcal{H}) \setminus \mathcal{F}| \text{ binary clauses} \end{cases}$$

Inclusion Constraint. Elements of $\mathcal{F}$ that are in F must also be in G:

$$F \subseteq G \iff_{enc} \forall x \in \mathcal{F}, \ x_{\mathcal{F}} \rightarrow x_{\mathcal{G}} \qquad |\mathcal{F}| \text{ binary clauses}$$

Difference Constraint. After reduction, F and H have the same support:

- for the elements of $\mathcal{F} \cap \mathcal{G} \cap \mathcal{H}$: such elements are in H if and only if they are in F and not in G;
- for the elements of $\mathcal{F} \setminus \mathcal{G}$: they are in H if and only if they are in F.

$$H = F \setminus G \quad \Leftrightarrow_{enc}$$

$$\begin{cases} \forall x \in \mathcal{F} \cap \mathcal{G} \cap \mathcal{H},\ x_F \wedge \neg x_G \leftrightarrow x_H \\ \forall x \in \mathcal{F} \setminus \mathcal{G},\ x_F \leftrightarrow x_H \end{cases} \quad \begin{array}{l} |\mathcal{F} \cap \mathcal{G} \cap \mathcal{H}|\ \text{ternary clauses} \\ +2.|\mathcal{F} \cap \mathcal{G} \cap \mathcal{H}|\ \text{binary clauses} \\ 2.|\mathcal{F} \setminus \mathcal{G}|\ \text{binary clauses} \end{array}$$

Multi-union Constraint. The multi-union constraint $H = \bigcup_{i=1}^{n} F_i$ is equivalent to the $n-1$ ternary constraints: $F_{1,2} = F_1 \cap F_2$, $F_{1,2,3} = F_{1,2} \cap F_3$, ... It is not only a short-hand, but it also significantly reduces the number of variables (only variables for H are required, not for each set $F_{1,2,...}$) and generated clauses. Indeed, elements of $\bigcap_{i=1}^{n} \mathcal{F}_i$ are considered once in the multi-union constraint whereas they are considered $n-1$ times in the corresponding $n-1$ binary union constraints. the set $\{1, \ldots, n\}$.

$$H = \bigcup_{i=1}^{n} F_i \Leftrightarrow_{enc} \begin{cases} \forall i \in N,\ \forall x \in F_i,\ x_{F_i} \to x_H \\ \forall x \in \mathcal{H},\ x_H \to \bigvee_{i \in N, x \in F_i} x_{F_i} \end{cases} \quad \begin{array}{l} \sum_{i=1}^{n} |\mathcal{F}i|\ \text{binary clauses} \\ |\mathcal{H}|\ m\text{-ary clauses}\ (m \leq n) \end{array}$$

Multi-intersection Constraint. Similarly, we define the multi-intersection constraints. As for the multi-union, the advantage is the gain of clauses and variables in the generated SAT instance:

$$H = \bigcap_{i \in N} F_i \quad \Leftrightarrow_{enc}$$

$$\begin{cases} \forall x \in \mathcal{H},\ \bigwedge_{i=1}^{n} x_{F_i} \leftrightarrow x_H \\ \forall x \in (\bigcap_{i=1}^{n} \mathcal{F}_i) \setminus \mathcal{H},\ \bigvee_{i \in N}(\neg x_{F_i}) \end{cases} \quad \begin{array}{l} 2.|\mathcal{H}|\ (n+1)\text{-ary clauses} \\ |\bigcap_{i \in N} \mathcal{F}_i \setminus \mathcal{H}|\ n\text{-ary clauses} \end{array}$$

Cardinality Constraint. This constraint has been studied for the encoding of global constraints (see e.g., [4]). The very intuitive encoding is quite simple but the generated clauses are too large. A more efficient encoding is based on the unary representation of integers (an integer $k \in [0..n]$ is represented by 1 k times followed by 0 $n-k$ times). We re-use this encoding [4] that we have chosen for the unit clauses it generates, and thus, the simplifications that can be achieved in the SAT instances. Consider the set G over the support $\mathcal{G}$ of size n, then the set constraint $|G| = k$ generates: $n + \sum_{i=1}^{n} 2u_i^n(\lfloor \frac{u_i^n}{2} \rfloor + 1)(\lceil \frac{u_i^n}{2} \rceil + 1) - (\frac{u_i^n}{2} + 1)$ clauses and $\sum_{i=1}^{n} u_i^n$ variables. with $u_n^n = 1, u_1^n = n$ and $u_i^n = u_{2i-1}^n + 2u_{2i}^n + u_{2i+1}^n$. The *cardinality_le* constraint can similarly be generated.

5 Application to the Nqueen Problem

Practically, the $\Rightarrow_{red}$ rules have been implemented as Constraint Handling Rules (CHR [7]), and the $\Leftrightarrow_{enc}$ rules with C++. To illustrate our approach, we have chosen the nqueen problem for various reasons: it is not well suited for SAT solvers; it scales well; it can be modeled in various ways with sets. We first give an intuitive model and then a more efficient model of the nqueen problem.

nqB: Model with the Board as Universe. The variables are the following:

- Universe: $\mathcal{U} = \{x_{1,1}, \ldots, x_{n,n}\}$, i.e., the set of cells of a $n \times n$ board
- Rows: $\forall i \in [1..n], set(R_i, \{x_{i,1}, \ldots, x_{i,n}\})$
- Columns: $\forall i \in [1..n], set(C_i, \{x_{1,i}, \ldots, x_{n,i}\})$
- $2.n - 3$ East-West diagonals: $set(D_1, \{x_{1,2}, x_{2,1}\}), set(D_2, \{x_{1,3}, x_{2,2}, x_{3,1}\})$, $\ldots, set(D_{2.n-3}, \{x_{n-1,n}, x_{n,n-1}\})$
- $2.n - 3$ West-East diagonals: $set(D_{2.n-2}, \{x_{n-1,2}, x_{n,1}\}), \ldots, set(D_{4.n-6}, \{x_{1,n-1}, x_{2,n}\})$
- the set of n queens: $set(Q, \{x_{1,1}, \ldots, x_{n,n}\})$
- the n queens: $\forall i \in [1..n], Element(q(i), \{x_{1,1}, \ldots, x_{n,n}\})$

The constraints are:

- Q is of size n: $|Q| = n$
- the n queens are in Q: $\forall i \in [1..n], q(i) \in Q$
- queen i is on row i: $\forall i \in [1..n], q_i \in R_i$
- one and only one queen per column: $\forall i \in [1..n], set(CQ_i, \{x_{1,1}, \ldots, x_{n,n}\})$, $CQ_i = C_i \cap Q, |CQ_i| = 1$
- at most one queen per diagonal: $\forall i \in [1..4.n - 6], set(DQ_i, \{x_{1,1}, \ldots, x_{n,n}\})$, $DQ_i = D_i \cap Q, |DQ_i| < 2$

Note that the support of each CQ_i (resp. DQ_i) could have been set to the support of C_i (resp. D_i). However, one does not have to care about this when modeling since the $\Rightarrow_{red}$ rules will reduce these supports. The solutions are contained in Q: each element of Q is a queen, i.e., a cell of the board.

nqQ: Model with the Queens as Universe. Since the encoding is very correlated to the size of the support, we propose another model where the universe is much smaller, i.e., the set of n queens:

- Universe: $\mathcal{Q} = \{q_1, \ldots, q_n\}$, i.e., the n queens to be placed on a $n \times n$ board;
- Rows: $\forall i \in [1..n], set(R_i, \{q_i\})$; each row i is over queen i;
- the set of queens and columns are defined as above, but over the support $\mathcal{Q}$;
- each cell $C_{i,j}$ is defined as the intersection of row R_i and column C_j: $\forall i, j \in [1..n], set(C_{i,j}, \mathcal{Q}), C_{i,j} = R_i \cap C_j$;
- the $4.n - 6$ diagonals are defined by unions of cells: $set(D_1, \mathcal{Q}), D_1 = C_{1,2} \cup C_{2,1}, set(D_2, \mathcal{Q}), D_2 = C_{1,3} \cup C_{2,2} \cup C_{3,1}, \ldots$
- Q is of size n: $|Q| = n$
- to enforce one queen per row: $\forall i \in [1..n], |R_i| = 1$;
- one and only one queen per column: $\forall i \in [1..n], |C_i| = 1$;
- a different queen on each column: $Q = \bigcup_{i=1}^{n} C_i$ (or, $\forall i, j \in [1..n], set(CC_{i,j}, \mathcal{Q}), CC_{i,j} = C_i \cap C_j, |CC_{i,j}| = 0$);
- atmost one queen per diagonal: $\forall i \in [1..4.n - 6], |D_i| < 2$.

Table 1. Experimental results

q	Model with the Board as universe								
	$\Leftrightarrow_{enc}$				$\Rightarrow_{red} + \Leftrightarrow_{enc}$				
	var	cl	time $\Leftrightarrow_{enc}$	minisat	time $\Rightarrow_{red}$	var	cl	time $\Leftrightarrow_{enc}$	minisat
5	3 696	22 749	0,07	0,01	0,02	564	2 055	0,01	0,00
10	42 956	635 874	2,19	0,38	0,06	2 876	18 654	0,09	0,03
15	170 611	4 652 164	16,64	2,94	0,21	7 243	75 413	0,25	0,13
20	449 116	19 336 154	56,90	18,07	0,42	13 924	212 850	0,68	0,65
25	942 141	58 637 309	183,67	88,12	0,75	22 977	486 081	1,61	2,92
30	1 715 296	145 441 474	485,82		0,78	34 392	965 086	3,77	8,37
35					1,93	48 486	1 735 499	6,70	18,54
40					2,79	65 316	2 897 434	12,56	43,23
45					4,00	84 646	4 565 419	19,65	89,72
50					5,32	106 928	6 870 358	32,06	166,80
55					7,16	131 783	9 956 493	50,23	303,35
60					9,18	159 188	13 983 778	76,83	514,04
65									
70									
75									
80									
85									
90									
95									
100									

q	Model with the queens as universe								
	$\Leftrightarrow_{enc}$				$\Rightarrow_{red} + \Leftrightarrow_{enc}$				
	var	cl	time $\Leftrightarrow_{enc}$	minisat	time $\Rightarrow_{red}$	var	cl	time $\Leftrightarrow_{enc}$	minisat
5	475	1 486	0,01	0,00	0,03	267	796	0,01	0,00
10	3 000	11 166	0,07	0,00	0,05	1 332	4 936	0,03	0,01
15	8 585	35 566	0,16	0,02	0,19	3 347	14 476	0,07	0,01
20	18 300	81 326	0,37	0,05	0,34	6 428	31 398	0,13	0,04
25	32 835	154 326	0,72	0,09	0,61	10 623	57 548	0,21	0,07
30	52 870	260 426	1,20	0,16	0,92	15 918	94 648	0,34	0,09
35	79 665	406 646	1,84	0,26	1,37	22 439	144 700	0,50	0,16
40	114 000	599 046	2,81	0,35	2,09	30 264	209 610	0,69	0,24
45	156 335	843 046	3,79	0,51	3,02	39 339	291 020	0,94	0,32
50	207 420	1 144 646	5,20	0,68	4,13	49 664	390 680	1,24	0,45
55	268 005	1 509 846	6,87	0,96	5,42	61 239	510 340	1,59	0,58
60	338 840	1 944 646	8,80	1,24	7,08	74 064	651 750	2,00	0,73
65	420 995	2 455 686	11,17	1,67	9,45	88 207	816 796	2,54	0,95
70	516 330	3 051 186	13,63	1,95	11,91	103 962	1 007 816	3,15	1,18
75	624 415	3 734 786	16,97	4,54	16,70	121 117	1 226 136	3,77	1,44
80	746 000	4 512 486	20,32	2,81	20,80	139 672	1 473 506	4,43	1,77
85	881 835	5 390 286	24,20	3,74	25,90	159 627	1 751 676	5,37	2,12
90	1 032 670	6 374 186	28,71	4,16	32,00	180 982	2 062 396	6,10	2,58
95	1 199 255	7 470 186	33,82	9,57	37,80	203 737	2 407 416	7,09	3,06
100	1 382 340	8 684 286	45,03	17,63	40,50	227 892	2 788 486	8,47	3,46
105	1 582 675	10 022 486	46,08	8,54	52,00	253 447	3 207 356	9,55	4,22
110	1 801 010	11 490 786	53,00	142,33	61,00	280 402	3 665 776	11,15	4,68
115	2 038 095	13 095 186	60,13	61,73	70,80	308 757	4 165 496	12,72	5,46
120	2 294 680	14 841 686	69,36		82,00	338 512	4 708 266	14,39	6,12
150	4 302 080	28 673 566	132,96		185,00	550 732	8 975 516	26,40	12,98
200	9 735 680	67 150 566	316,90		525,00				

Interpretation of the results: if cell $C_{i,j} = \{q_k\}$, then queen q_k is in $i \times j$, else $C_{i,j} = \emptyset$ and there is no queen in $i \times j$.

As said before, our goal is not to compete with arithmetic solvers or set solvers, but to be able to declaratively, expressively, and error-prone model problems into SAT. Table 5 presents the results for the two models (model with the board as universe and model with the queens as universe). Column "q" represents the queens number and others columns represent the number of variables (var) and clauses (cl) for the generated SAT instance, the encoding time (time $\Leftrightarrow_{enc}$), the reduction time (time $\Rightarrow_{red}$) and the solving time by the Minisat solver [6] (minisat). When have limited the running time to 600 seconds for each combination of processes. No result is written if this value is reached. When only the Minisat column is empty this means that the instance exceed the memory size (4GB).

We can observe that the reduction rules $\Rightarrow_{red}$ permit to significantly reduce the size of the SAT instances. Thereby, instances which are unsolvable (due to the size) before reduction are now solved by MiniSat (q=30 for model nqB and q=120 for model nqQ). This result shows that contrary to some reduction rules such as breaking symmetry [17], our reduction rules do not make the search more difficult. Finally we can also note that the cumulative running time (encoding+resolution or reduction+encoding+resolution) is better when reduction is applied: always for the nqB model and from q=30 for the nqQ model.

We have tried our technique on various problems (e.g., Social Golfer Problem [11], Sudoku, Car-sequencing) and the SAT instances which are automatically generated have a complexity similar to the complexity of improved hand-written SAT formulations, and their solving with a SAT solver (in our case Minisat) is efficient compared to other SAT approaches.

6   Conclusion

We have presented a technique for encoding set constraints into SAT: the modeling process is achieved using some very declarative and expressive set constraints; they are then reduce by our $\Rightarrow_{red}$ rules before being automatically converted ($\Leftrightarrow_{enc}$) into SAT variables and clauses. We have illustrated our approach on the nqueen problem and shown some good results with the application of reduction and encoding rules. The advantages of our technique are the following:

- the modeling process is simple, declarative, expressive, and readable. Moreover, it is solver independent and independent from CSP or SAT solvers;
- the technique is less error-prone than hand-written SAT encodings;
- the SAT instances which are automatically generated are smaller in terms of number of variables and clauses;
- finally, with respect to solving time, adding reduction process permits to reduce the cumulative running time (reduction+encoding+resolution);
- the generated SAT instances also appeared to be well-suited for Minisat.

In the future, we plan to use our set constraints encoding for formalizing finite domain variables. We also plan to combine set constraints with arithmetic constraints, and we want to define the corresponding combining SAT encoding. To this end, we will need to add some new constraints and to complete our $\Leftrightarrow_{enc}$ and $\Rightarrow_{red}$ rules.

References

1. Choco, http://www.emn.fr/z-info/choco-solver/
2. Apt, K.: Principles of Constraint Programming. Cambridge University Press (2003)
3. Bacchus, F.: Gac via unit propagation. In: Bessière, C. (ed.) CP 2007. LNCS, vol. 4741, pp. 133–147. Springer, Heidelberg (2007)
4. Bailleux, O., Boufkhad, Y.: Efficient CNF encoding of boolean cardinality constraints. In: Rossi, F. (ed.) CP 2003. LNCS, vol. 2833, pp. 108–122. Springer, Heidelberg (2003)
5. Bessière, C., Hebrard, E., Walsh, T.: Local consistencies in SAT. In: Giunchiglia, E., Tacchella, A. (eds.) SAT 2003. LNCS, vol. 2919, pp. 299–314. Springer, Heidelberg (2004)
6. Eén, N., Sörensson, N.: An extensible sat-solver. In: Giunchiglia, E., Tacchella, A. (eds.) SAT 2003. LNCS, vol. 2919, pp. 502–518. Springer, Heidelberg (2004)
7. Frühwirth, T.: Constraint Handling Rules. Cambridge University Press (2009)
8. Garey, M.R., Johnson, D.S.: Computers and Intractability, A Guide to the Theory of NP-Completeness. W.H. Freeman & Company (1979)
9. Gent, I., Lynce, I.: A sat encoding for the social golfer problem. In: IJCAI 2005 Workshop on Modelling and Solving Problems with Constraints (2005)
10. Gervet, C.: Conjunto: Constraint propagation over set constraints with finite set domain variables. In: Proc. of ICLP 1994, p. 733. MIT Press (1994)
11. Lardeux, F., Monfroy, E., Crawford, B., Soto, R.: Set constraint model and automated encoding into sat: Application to the social golfer problem. Submitted to Annals of Operations Research
12. Lardeux, F., Monfroy, E., Saubion, F., Crawford, B., Castro, C.: SAT encoding and CSP reduction for interconnected alldiff constraints. In: Aguirre, A.H., Borja, R.M., Garciá, C.A.R. (eds.) MICAI 2009. LNCS, vol. 5845, pp. 360–371. Springer, Heidelberg (2009)
13. Legeard, B., Legros, E.: Short overview of the CLPS system. In: Małuszyński, J., Wirsing, M. (eds.) PLILP 1991. LNCS, vol. 528, pp. 431–433. Springer, Heidelberg (1991)
14. Mackworth, A.: Encyclopedia on Artificial Intelligence, chapter Constraint Satisfaction. John Wiley (1987)
15. Monfroy, E.: A coordination-based chaotic iteration algorithm for constraint propagation. In: Proc of ACM SAC 2000 (1), pp. 262–269. ACM (2000)
16. Monfroy, E., Saubion, F., Lambert, T.: On hybridization of local search and constraint propagation. In: Demoen, B., Lifschitz, V. (eds.) ICLP 2004. LNCS, vol. 3132, pp. 299–313. Springer, Heidelberg (2004)
17. Prestwich, S., Roli, A.: Symmetry breaking and local search spaces. In: Barták, R., Milano, M. (eds.) CPAIOR 2005. LNCS, vol. 3524, pp. 273–287. Springer, Heidelberg (2005)
18. Rossi, F., van Beek, P., Walsh, T. (eds.): Handbook of Constraint Programming. Elsevier (2006)
19. Triska, M., Musliu, N.: An improved sat formulation for the social golfer problem. Annals of Operations Research 194(1), 427–438 (2012)

A Mathematical Hierarchy of Sudoku Puzzles and Its Computation by Boolean Gröbner Bases

Shutaro Inoue and Yosuke Sato

Tokyo University of Science, 1-3, Kagurazaka, Shinjuku-ku, Tokyo, Japan
{sinoue,ysato}@rs.kagu.tus.ac.jp

Abstract. Sudoku, which is one of the most popular puzzles in the world, can be considered as a kind of combinatorial problem. Considering a Sudoku puzzle as a singleton set constraint, we define a purely mathematical hierarchy of Sudoku puzzles in terms of a Boolean polynomial ring. We also introduce a sufficiently practical symbolic computation method using Boolean Gröbner bases to determine the hierarchy level of a given Sudoku puzzle. According to our experiments through our implementation, there exists a strong positive correlation between our hierarchy and the levels of difficulty of Sudoku puzzles usually assigned by a heuristic analysis. Our mathematical hierarchy would be a universal tool which ensures the mathematical correctness of the level of a Sudoku puzzle given by a heuristic analysis.

Keywords: Sudoku, Boolean Gröbner Bases.

1 Introduction

Combinatorial problems are often reduced to solving constraints of sets. Such constraints are described as polynomial equations over Boolean rings of sets with some additional conditions of cardinality. Sudoku that is one of the most popular puzzles in the world is among such instances. Any Sudoku puzzle can be considered as a constraint of sets with an additional condition that each variable has to be a singleton, i.e., a set containing exactly one element. When we solve a Sudoku puzzle as a human, i.e., without a computer, we usually use several techniques such as naked pair/triple, hidden pair/triple, XY-wing, XY-chain, etc. Those techniques are categorized from the easiest level to the highest level. Most books and websites of Sudoku puzzles assign the level of difficulty to each puzzle, which is usually given by a heuristic analysis of the applicable techniques. Hence, it might happen that different analyses assign different levels of difficulty to one puzzle. Based on the theory of Boolean polynomial ring, we define a purely mathematical hierarchy of Sudoku puzzles. Our hierarchy reflect any generalization of the above techniques. We also introduce a sufficiently practical method to determine the hierarchy level of a given Sudoku puzzle, which is based on the computations of Boolean Gröbner bases [10]. Our method is implemented on the computer algebra system Risa/Asir [8]. (See also [4].) In order to see our hierarchy is effective, we computed the hierarchy levels for 735

G.A. Aranda-Corral et al. (Eds.): AISC 2014, LNAI 8884, pp. 88–98, 2014.

Sudoku puzzles in the series of Sudoku books [12] where they are categorized from the level 1 to 7 according to their levels of difficulty assigned by some heuristic analysis. Our computation experiments tell us that there exists a strong positive correlation between our hierarchy and the levels of difficulty assigned in the book. Some heuristic analysis sometimes assigns an improper level of difficulty for some puzzle. Our mathematical hierarchy would be a universal tool which ensures the mathematical correctness of the level of a Sudoku puzzle given by a heuristic analysis.

The paper is organized as follows. In section 2, we show how we can consider a Sudoku puzzle as a singleton set constraint and how we can translate it into a system of equations of a certain Boolean polynomial ring. We also give a minimum description of a Boolean polynomial ring. In section 3, we define a hierarchy of Sudoku puzzles. In section 4, we show some properties concerning the computation of minimal polynomials in a Boolean polynomial ring. These properties enable us to compute the hierarchy level of a given Sudoku puzzle. Section 5 contains some computation data we have obtained through our computation experiments. The reader is referred to [10] for Boolean Gröbner bases and related properties of Boolean rings.

2   Sudoku Puzzle as a Singleton Set Constraint

We begin with a quick review of Boolean polynomial rings.

Definition 1. A commutative ring $\mathbf{B}$ with an identity 1 is called *a Boolean ring* if every element a of $\mathbf{B}$ is idempotent, i.e., $a^2 = a$.

$(\mathbf{B}, \vee, \wedge, \neg)$ becomes a Boolean algebra with the Boolean operations $\vee, \wedge, \neg$ defined by $a \vee b = a + b + a \cdot b, a \wedge b = a \cdot b, \neg a = 1 + a$. Conversely, for a Boolean algebra $(\mathbf{B}, \vee, \wedge, \neg)$, if we define $+$ and $\cdot$ by $a + b = (\neg a \wedge b) \vee (a \wedge \neg b)$ and $a \cdot b = a \wedge b$, $(\mathbf{B}, +, \cdot)$ becomes a Boolean ring. Note that $+$ is nothing but an exclusive OR operator. Since $-a = a$ in a Boolean ring, we do not need to use the symbol '$-$', however, we also use $-$ when we want to stress its meaning.

Example 1. Let S be an arbitrary set and $\mathcal{P}(S)$ be its power set, i.e., the family of all subsets of S. Then, $(\mathcal{P}(S), \vee, \wedge, \neg)$ becomes a Boolean algebra with the operations $\vee, \wedge, \neg$ as union, intersection and the complement of S respectively.

Definition 2. Let $\mathbf{B}$ be a Boolean ring. A quotient ring $\mathbf{B}[X_1, \ldots, X_n]/\langle X_1^2 - X_1, \ldots, X_n^2 - X_n \rangle$ modulo an ideal $\langle X_1^2 - X_1, \ldots, X_n^2 - X_n \rangle$ becomes a Boolean ring. It is called *a Boolean polynomial ring* and denoted by $\mathbf{B}(X_1, \ldots, X_n)$, its element is called *a Boolean polynomial*.

Note that a Boolean polynomial of $\mathbf{B}(X_1, \ldots, X_n)$ is uniquely represented by a polynomial of $\mathbf{B}[X_1, \ldots, X_n]$ that has at most degree 1 for each variable X_i. In what follows, we identify a Boolean polynomial with such a representation.

Multiple variables such as $X_1, \ldots, X_n$ or $Y_1, \ldots, Y_m$ are abbreviated to $\bar{X}$ or $\bar{Y}$ respectively. Lower small roman letters such as a, b, c are usually used for elements of a Boolean ring **B**. The symbol $\bar{a}$ denotes an m-tuple of elements of **B** for some m.

Definition 3. Let I be an ideal of $\mathbf{B}(\bar{X})$. For a subset S of $\mathbf{B}^n$, $V_S(I)$ denotes a subset $\{\bar{a} \in S | \forall f \in I f(\bar{a}) = 0\}$. When $S = \mathbf{B}^n$, $V_S(I)$ is simply denoted by $V(I)$ and called a *variety* of I. We say I is *satisfiable* in S if $V_S(I)$ is not empty. When $S = \mathbf{B}^n$, we simply say I is *satisfiable*.

We first show how a Sudoku puzzle is considered as a singleton set constraint and presented as a system of equations of a Boolean polynomial ring.

4				9				
3					1	8		
			5					
			5	8				
		2	9					
					1	7		
		6					5	
					7			
			2					9

Consider the above Sudoku puzzle. We associate a variable X_{ij} for each grid at the i-th row and the j-th column. This puzzle can be considered as a set constraint where each variable should be assigned a singleton from 9 candidates $\{1\}, \{2\}, \{3\}, \{4\}, \{5\}, \{6\}, \{7\}, \{8\}$ and $\{9\}$ so that any distinct two variables which lie on a same row, column or block must be assigned different singletons. 17 variables are assigned singletons $X_{11} = \{4\}, X_{15} = \{9\}, \ldots, X_{99} = \{9\}$ as the initial conditions. This constraint is translated into a system of equations of the Boolean polynomial ring $\mathbf{B}(X_{11}, X_{12}, \ldots, X_{99}) = \mathbf{B}(\bar{X})$ with $\mathbf{B} = \mathcal{P}(\{1, 2, \ldots, 9\})$ as follows:

(1) $X_{11} = \{4\}, X_{15} = \{9\}, \ldots, X_{99} = \{9\}$.
(2) $X_{ij}X_{i'j'} = 0(= \emptyset)$ for each pair of distinct variables $X_{ij}, X_{i'j'}$ which lie on a same row, column or block.
(3) $\sum_{(i,j) \in A} X_{ij} = 1(= \{1, 2, \ldots, 9\})$ where A is a set of indices lying on a same row, column or block. (There are 27 such A's.)

This puzzle is nothing but solving the above equations with a strong restriction that each variable must be a singleton. Let I be the ideal of $\mathbf{B}(\bar{X})$ generated by the corresponding polynomials of (1),(2) and (3), we call I the *corresponding ideal* of the puzzle. Let $Sing$ denote the subset of $\mathbf{B}^{81}$ defined by $Sing = \{(a_1, a_2, \ldots, a_{81}) \in \mathbf{B}^{81} |$ each a_i is a singleton $\}$. Then the puzzle is nothing but obtaining $V_{Sing}(I)$. We can easily compute $V(I)$ by the computation of a stratified Boolean Gröbner basis of I, however we have to do something more for the computation of $V_{Sing}(I)$.

In the rest of the paper, **B** denotes the Boolean ring $\mathcal{P}(\{1, 2, \ldots, 9\})$, $\bar{X}$ denotes 81 variables $X_{11}, \ldots, X_{99}$.

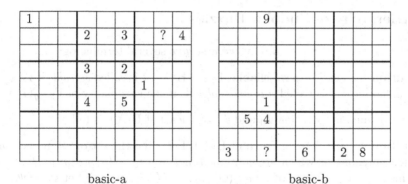

basic-a   basic-b

The above two pictures illustrate two types of typical examples of the basic level strategies of Sudoku solving. In the left example the variable X_{28} must be $\{1\}$, in the right example the variable X_{93} must be $\{7\}$.

The corresponding ideal contains a unary polynomial $\{1,2,3,4\}X_{28}+\{1\}$ (but not $X_{28}+\{1\}$) for the left example, $(1+\{7\})X_{93}$(but not $X_{93}+\{7\}$) for the right example. Note that the equation $\{1,2,3,4\}X_{28}+\{1\}=0$ has a solution $X_{28}=\{1\}\cup p$ with any subset p of $\{5,6,7,8,9\}$, but the singleton solution is just $X_{28}=\{1\}$. The equation $(1+\{7\})X_{93}=0$ has another solution $X_{93}=\emptyset$. These observations lead us to the following definitions.

Definition 4. A unary Boolean polynomial $f(X)\in\mathbf{B}(X)$ is called
  (i) a *solution polynomial* if it has the following form
    $f(X)=X+\{s\}$,
  (ii) a *semi-solution polynomial of type-a* if it has the following form
    $f(X)=\{s_1,s_2,\ldots,s_l\}X+\{s_i\}$ for some $i=1,\ldots,l$
 (iii) a *semi-solution polynomial of type-b* if it has the following form
    $f(X)=(1+\{s\})X$
 (iv) a *contradiction polynomial* if it has the following form
    $f(X)=a$, $f(X)=X$ or $f(X)=aX+b$ for some non-zero constant a,b
     such that the cardinality of b is greater than 1.
The polynomials $X+\{s_i\}$ and $X+\{s\}$ are called the *associated solution polynomial* of the semi-solution polynomial of (ii) and (iii) respectively. The associated solution polynomial of a semi-solution polynomial h is denoted $asp(h)$.

Obviously we have the following property.

Lemma 5. Let I be an ideal of $\mathbf{B}(\bar{X})$ generated by the corresponding polynomials of (1),(2) and (3). If I contains a semi-solution polynomial $f(X)$, let $X+\{s\}$ be its associated solution polynomial then the singleton solution of X must be $\{s\}$, more precisely $V_{Sing}(I)=V_{Sing}(I+\langle X+\{s\}\rangle)$. If I contains a contradiction polynomial then $V_{Sing}(I)=\emptyset$.

3   Hierarchy of Sudoku Puzzles

In order to describe our hierarchy, we first give several terminologies.

Definition 6. Let I be a satisfiable ideal of $\mathbf{B}(\bar{X})$. I is called *solvable* if $V_{Sing}(I)$ $\neq \emptyset$, furthermore I is called *uniquely solvable* if the cardinality of $V_{Sing}(I)$ is 1.

We first define an operation Ψ_0 and Ψ_0^* on ideals of $\mathbf{B}(\bar{X})$ as follows.

Definition 7. Let I be an ideal of $\mathbf{B}(\bar{X})$. Let P be the set of all unary semi-solution polynomials which are contained in I and $Q = \{asp(h)|h \in P\}$, $\Psi_0(I)$ is defined as an ideal sum of I and $\langle Q \rangle$, i.e., $\Psi_0(I) = I + \langle Q \rangle$. Let $\Psi_0^1 = \Psi_0$ and $\Psi_0^{k+1}(I) = \Psi_0(\Psi_0^k(I))$ for each $k = 1, \dots$. Since $\Psi_0^1(I), \Psi_0^2(I), \dots$ are increasing sequence and $\mathbf{B}(\bar{X})$ is a finite Boolean ring, there exists m such that $\Psi_0^{m+1}(I) = \Psi_0^m(I)$. Let m be the least one and set $\Psi_0^*(I) = \Psi_0^m(I)$. $\Psi_0^*(I)$ is called the *basic closure* of I. An ideal I is said to be *basically closed* if $\Psi_0^*(I) = I$. Note that $\Psi_0^*(I)$ is basically closed.

Definition 8. For a uniquely solvable ideal I, if $\Psi_0^*(I)$ is a maximal ideal, i.e., it is generated by 81 solution polynomials $X_{11} + \{s_{11}\}, \dots, X_{99} + \{s_{99}\}$, then I is said to be *basic solvable*. For a basic solvable ideal I, its *basic rank* is the least m such that $\Psi_0^*(I) = \Psi_0^m(I)$ denoted by b-rank(I).

If any uniquely solvable ideal is also basic solvable, then we would be able to give a mathematical hierarchy of Sudoku puzzles by using the basic rank defined above. Unfortunately, the situation is not so simple. There exist many uniquely solvable ideals which are not basic solvable but can be handled by more advanced strategies of Sudoku solving. For making a hierarchy for such tough ideals, we have to do something more.

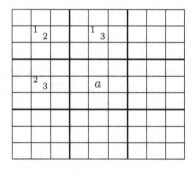

XY-wing   XY-chain

The above two pictures illustrate two types of typical examples of well-known advanced strategies of Sudoku solving. In the picture there are two numbers located in a same grid. For example 1 and 2 are located in the grid of the second line and the second column. It means that there are only two possible singleton solutions $\{1\}, \{2\}$ for the variable X_{22}. The XY-wing strategy tells us that $a \neq 3$ in the left picture, the XY-chain strategy tells us that $a \neq 1, b \neq 2, c \neq 3$ and $d \neq 4$ in the right picture. The given situation is generalized as a unary polynomial equation $(1 + \{1,2\})X_{22} = 0$. That is the polynomial $(1 + \{1,2\})X_{22}$ is contained in the corresponding ideal. For the left example, let the initial polynomials of (1) of a Sudoku puzzle be $(1 + \{1,2\})X_{22}, (1 + \{2,3\})X_{52}$ and $(1 + \{1,3\})X_{25}$, then consider the corresponding ideal I. If I contains the polynomial $\{3\}X_{55}$, we can detect $a \neq 3$. Unfortunately I does not contain it, actually I contains no unary polynomials. On the other hand, the ideal $I + \langle X_{55} + \{3\}\rangle$ contains two semi-solution polynomials $(1 + \{1\})X_{25}$ and $(1 + \{2\})X_{52}$. If we further put their associated solution polynomials, then $I + \langle X_{55} + \{3\}, X_{25} + \{1\}, X_{52} + \{2\}\rangle$ contains the contradiction polynomial X_{22}. Let the initial polynomials be $(1+\{1,2\})X_{22}, (1+\{2,3\})X_{52}, (1+\{3,4\})X_{55}$, and $(1+\{1,4\})X_{25}$ for the right example. In this case there are many variables whose candidate can be eliminated. Consider the variable X_{24} for example. $X_{24} \neq 1$ is also detected as follows. The ideal $I + \langle X_{24} + \{1\}\rangle$ contains two semi-solution polynomials $(1 + \{2\})X_{22}$ and $(1 + \{4\})X_{25}$. The ideal $I + \langle X_{24} + \{1\}, X_{22} + \{2\}, X_{25} + \{4\}\rangle$ contains two semi-solution polynomials $(1 + \{3\})X_{52}$ and $(1 + \{3\})X_{55}$. Finally, the ideal $I + \langle X_{24} + \{1\}, X_{22} + \{2\}, X_{25} + \{4\}, X_{52} + \{3\}, X_{55} + \{3\}\rangle$ contains the contradiction polynomial $\{3\}$. The elimination for the other variables can be similarly detected. In the left example we need one step of manipulation of refinement of an ideal by the associated solution polynomials to get a contradiction polynomial, in the right example we need two steps of manipulations. In general the human strategy of XY-chain is considered more advanced than XY-wing. These observation lead us to the following definition of a further hierarchy for uniquely solvable but not basic solvable ideals.

Definition 9. Let I be an ideal of $\mathbf{B}(\bar{X})$. For each variable X, a solution polynomial $X + \{s\}$ is said to be *basic refutable* w.r.t. I if $\Psi_0^*(I + \langle X + \{s\}\rangle)$ contains a contradiction polynomial. When $X + \{s\}$ is refutable in I, the least m such that $\Psi_0^m(I + \langle X + \{s\}\rangle)$ contains a contradiction polynomial is called the *basic refutable rank*(br-rank in short) of $X + \{s\}$ w.r.t. I and denoted br-rank$(X + \{s\}, I)$ (abbreviated br-rank$(X + \{s\})$ when I is clear from context).

Using these terminologies, we define operations $\Psi_1, \Psi_1^*, \Psi_2, \Psi_2^*, \ldots$ and Ψ_∞^* on ideals of $\mathbf{B}(\bar{X})$ as follows.

Definition 10. Fix a natural number k. For an arbitrary ideal J of $\mathbf{B}(\bar{X})$. Define the set $BR_k(J)$ of unary polynomials by $BR_k(J) = \{\{s\}X | X + \{s\}$ is a basic refutable polynomial w.r.t. J such that br-rank$(X+\{s\}) \leq k\}$. For an ideal I of $\mathbf{B}(\bar{X})$, set $\Psi_k(I) = \Psi_0^*(I + \langle BR_k(I)\rangle)$. Let $\Psi_k^1 = \Psi_k$ and $\Psi_k^{i+1}(I) = \Psi_k(\Psi_k^i(I))$ for each $k = 1, 2, \ldots$. Since $\Psi_k^1(I), \Psi_k^2(I), \ldots$ are increasing sequence, there exists a natural number m such that $\Psi_k^m(I) = \Psi_k^{m+1}(I) = \cdots$. For such an m set $\Psi_k^*(I) = \Psi_k^m$. Since $\Psi_1^*(I), \Psi_2^*(I), \ldots$ are also increasing sequence, there exists l such that $\Psi_l^*(I) = \Psi_{l+1}^*(I) = \cdots$. For such an l set $\Psi_\infty^*(I) = \Psi_l^*$. $\Psi_\infty^*(I)$ is called the *strategy closure* of I. An ideal I is said to be *strategy closed* if $\Psi_\infty^*(I) = I$.

Definition 11. Let I be a uniquely solvable but not basic solvable ideal. If $\Psi_\infty^*(I)$ is a maximal ideal, we say I is *strategy solvable*. For an strategy solvable ideal I the least k such that $\Psi_k^*(I) = \Psi_\infty^*(I)$ is called the *strategy rank* of I and denoted s-rank(I). We also assign s-rank 0 to any basic solvable ideal. If $\Psi_\infty^*(I)$ is not a maximal ideal, we assign s-rank ∞ to I.

4   Computation of Minimal Polynomials

In the previous section, we define a hierarchy of uniquely solvable ideals. For the computation of both ranks, we have to compute all the unary polynomials contained in a given ideal. In this section we show Boolean Gröbner bases are ideal tools for such computations.

Definition 12. Let I be a satisfiable ideal of a Boolean polynomial ring $\mathbf{B}(\bar{X})$. For a non-constant Boolean polynomial f of $\mathbf{B}(\bar{X})$, the set $\{p(Z) \in \mathbf{B}(Z) | p(f) \in I\}$ forms an ideal in a Boolean polynomial ring $\mathbf{B}(Z)$, where Z is a new variable. The Boolean polynomial $h(Z)$ which generates this ideal is called the *minimal polynomial* of f w.r.t. I. (Note that any finitely generated ideal in a Boolean ring is principal.) Such a Boolean polynomial is uniquely determined. Since we can use any variable Z, we denote such a unary Boolean polynomial simply by $MinPoly_{f,I}$. When I is clear from the context, we simply write $MinPoly_f$. When f is a variable X_i among $\bar{X}$, we always assume that $MinPoly_f$ is a polynomial of X_i.

Theorem 13. Let G be a Boolean Gröbner basis of a satisfiable ideal I in $\mathbf{B}(\bar{X})$. For a Boolean polynomial $f \in \mathbf{B}(\bar{X})$, let $\overline{f}^G = c_1 t_1 + \cdots + c_l t_l + d$, where $c_1, \ldots, c_l, d$ are elements of $\mathbf{B}$ and $t_1, \ldots, t_l$ are distinct Boolean terms which are not equal to 1. Then $MinPoly_{f,I}(Z) = (c_1 \vee \cdots \vee c_l + 1)(Z + d)$. (When $l = 0$, $c_1 \vee \cdots \vee c_l$ denotes 0.)

Proof. Note first that any unary Boolean polynomial is a linear polynomial. For arbitrary elements a, b of $\mathbf{B}$, $af + b \in I$ if and only if $\overline{af + b}^G = 0$. Since I is satisfiable, I does not contain a non-zero constant, so G does not contain a non-zero constant. Therefore $\overline{af + b}^G = a\overline{f}^G + b = a(c_1 t_1 + \cdots + c_l t_l + d) + b = ac_1 t_1 + \cdots + ac_l t_l + ad + b$, $af + b \in I \Leftrightarrow ac_1 = 0, \ldots, ac_l = 0$ and $ad + b = 0 \Leftrightarrow a(c_1 \vee \cdots \vee c_l) = 0$ and $b = ad \Leftrightarrow a = (c_1 \vee \cdots \vee c_l + 1)u, b = (c_1 \vee \cdots \vee c_l + 1)du$ for some element $u \in \mathbf{B} \Leftrightarrow aZ + b \in \langle (c_1 \vee \cdots \vee c_l + 1)Z + (c_1 \vee \cdots \vee c_l + 1)d \rangle = \langle (c_1 \vee \cdots \vee c_l + 1)(Z + d) \rangle$.   □

Note that we can use an arbitrary admissible term order in the above theorem. In case f is a single variable X_i, the next theorem shows that we do not even need monomial reductions, $MinPoly_{X_i}$ is essentially contained in the Gröbner basis.

Theorem 14. Let I be a satisfiable ideal in a Boolean polynomial ring $\mathbf{B}(\bar{X})$ and G be its stratified Boolean Gröbner basis w.r.t. an arbitrary admissible term order. Then for each variable X_i of $\bar{X}$, there exists a non-zero minimal polynomial of X_i w.r.t. I if and only if G contains a polynomial $g = aX_i + b_1t_1 + \cdots + b_lt_l + c$ with its leading term X_i, where $t_1, \ldots, t_l$ are distinct terms which are not equal to 1 and $b_1, \ldots, b_l, c$ are elements of $\mathbf{B}$ such that $a \neq b_1 \vee \cdots \vee b_l$. Moreover if such g exists, the minimal polynomial has the following form:

$$MinPoly_{X_i} = a(1 + b_1 \vee \cdots \vee b_l)X_i + c(1 + b_1 \vee \cdots \vee b_l).$$
(l and c could be 0. For $l = 0$, $b_1 \vee \cdots \vee b_l$ denotes 0).

Proof. If $X_i \notin LT(G) = \{LT(g) | g \in G\}$, then $\overline{X_i}^G = X_i$. In the previous theorem, let $f = X_i$, hence $l = 1$, $t_1 = X_i$, $c_1 = 1$ and $d = 0$. Therefore $Min_{X_i} = (1+1)(X_i + 0) = 0$. Otherwise, there exists $g \in G$ such that $LT(g) = X_i$. Let $g = aX_i + b_1t_1 + \cdots + b_lt_l + c$. $X_i \to_g (1+a)X_i + b_1t_1 + \cdots + b_lt_l + c$. Since G is reduced, $b_1t_1 + \cdots + b_lt_l + c$ is not reducible by $\to_G$. Since G is stratified, G does not contain any other element whose leading term is X_i, hence $(1 + a)X_i$ is not reducible by $\to_G$. Therefore $\overline{X_i}^G = (1 + a)X_i + b_1t_1 + \cdots + b_lt_l + c$. Note that $ab_1 = b_1, \ldots, ab_l = b_l$ and $ac = c$ since g is boolean closed. By the previous theorem, $MinPoly_{X_i} = ((1+a) \vee b_1 \vee \cdots \vee b_l + 1)(X_i + c) = ((1+a) + b_1 \vee \cdots \vee b_l + 1)(X_i + c) = (a + b_1 \vee \cdots \vee b_l)(X_i + c) = a(1 + b_1 \vee \cdots \vee b_l)X_i + c(1 + b_1 \vee \cdots \vee b_l)$. Obviously $MinPoly_{X_i} = 0$ if and only if $a = b_1 \vee \cdots \vee b_l$. $\qquad\square$

For a given Sudoku puzzle, we can compute the basic closure $\Psi_0^*(I)$ and the strategy closure $\Psi_\infty^*(I)$ of the corresponding ideal I together with its both ranks, i.e., the least l, m such that $\Psi_0^*(I) = \Psi_0^l(I)$ and $\Psi_\infty^*(I) = \Psi_m^*(I)$ by computing only Boolean Gröbner bases. From a Boolean Gröbner basis, we can directly obtain all unary minimal polynomials as is shown above.

5 Computation Data

Any computation of a Boolean Gröbner basis of an ideal of $\mathbf{B}(\bar{X})$ terminates within at most a few seconds in our implementation of Boolean Gröbner bases on the computer algebra system Risa/Asir. We have implemented a program to compute both ranks of a given Sudoku puzzle by the computations of Boolean Gröbner bases. Using our implementation we have computed both ranks for 735 Sudoku puzzles which are contained in the series of Sudoku books [12]. The books (named Basic,Middle,High,SuperHigh,Hard,SuperHard and UltraHard) are categorized from the easiest level 1 to the highest level 7. Each book has 105 Sudoku puzzles. While most Sudoku puzzle books contain only puzzles which are solvable by using some heuristic strategies, the series of books are famous for containing some tough puzzles which are not solvable by any known heuristic strategies. All puzzles in the book Basic and Middle are basic solvable, the average of b-rank is 16 for Basic and 20.5 for Middle. The book Basic contains 103 puzzles which can be solved using only a-type of semi-solution polynomials, meanwhile the book Middle contains 90 such puzzles. The other books contains puzzles which are not basic solvable. The following table contains the data we have obtained through our computation. Each grid contains the number of puzzles of the book of its line which has the s-rank of its column.

s-rank	0	1	2	3	4	5	∞
High	84	3	10	7	1	0	0
SuperHigh	58	9	22	12	4	0	0
Hard	39	15	21	17	8	4	0
SuperHard	17	13	32	24	19	1	0
UltraHard	11	15	22	21	21	9	6

The following pictures cite 6 examples of Sudoku puzzles we have computed. The first 5 puzzles are from the book UltraHard whose s-rank is ranged from 1 to 5. The last puzzle is the strategy closure of the tough puzzle introduced in [5]. That is the corresponding ideal of the picture is equal to $\Psi_\infty^*(I)$ where I is the corresponding ideal of the original puzzle of [5]. It is not strategy solvable.

s-rank 1

					6			
2	7	5	1				8	
4					8		9	
	3	7		4	5			
9			5			3		
6		9		1	8			
7		6						9
5				9	7	3	6	
	9							

s-rank 2

	6				5	3	4	
3		8		1				5
4			2				6	
5						6		
	9			7			1	
		6						9
3					1			4
9				6		7		1
	8	7	3				2	

s-rank 3

	8	6				4	3	
2			8		3			9
3				4				8
	2						1	
		1			7			
	3						8	
4			2					1
6			9		7			3
	1	3				2	5	

s-rank 4

			4	2		7	5	
	7				6		4	
1								6
5			4		3			8
	8			5			3	
4			9		6			5
7								1
	1			4			2	
		9	1		8	6		

s-rank 5

	6	2						3
1			8			5		
4				2	3	6		
	9							7
		3		7		5	6	
		4		5				
		7		3		4		
	4			5			9	
8			2					6

$\Psi_\infty^*(I)$

		5	3					
8				**5**			2	
	7			1		5		
4					5	3		
	1			7	**3**			6
		3	2				8	
	6		5					9
		4					3	
						9	7	

6 Conclusion and Remarks

In section 3, we have seen that some solution polynomials leading to a contradiction by XY-wing or XY-chain strategy are basic refutable. Similarly we can easily show that any solution polynomial leading to a contradiction by other existing Sudoku solving strategie is also basic refutable. Therefore, we can say that any sudoku puzzle which can be solved by existing Sudoku solving strategies is also strategy solvable defined in this paper.

We can also see there exists a strong positive correlation between our ranks and the difficalty levels of Sudoku puzzles in the books of [12]. Meanwhile, the hardest book contains 11 basic solvable puzzles. We think there exists some improper analysis for them. Though we can not simply say the higher rank puzzle is more difficult than the lower rank puzzle since there is no precise defenition of the "difficulty level" of Sudoku puzzles, our rank can be usuful for checking an analysis program.

The strategy closure $\Psi_\infty^*(I)$ is not equal to the corresponding ideal I of the puzzle of [5]. The original sudoku does not have 5 at the grid $(2,5)$ and 3 at the grid $(5,6)$. The same author introduced an improved puzzle in [6]. For its corresponding ideal I, we have $\Psi_\infty^*(I) = I$. There should be some mathematical structure describable in terms of our Boolean polynomial ring which makes clear the difference between those tough puzzles. We used only unary minimal polynomials for extending ideals. Consider the polynomial $(X \cup Y + 1)\{1, 2, 3\}) = 0$. Obviously it does not have a sigleton solution, however the ideal $\langle (X \cup Y + 1)\{1, 2, 3\}) \rangle$ does not contain any unary contradiction polynomial. If we use such a non-unary contradiction polynomial for the refutation process, we may enrich our hierarchy. The computation of the minimal polynomial of such a non-unary polynomial can be also done by the computations of Boolean Gröbner bases. But we need further monomial reductions by a Boolean Gröbner basis. This is actually a heavy computation. We are now implementing this computation for further expansion of our hierarchy.

Though we have not mentioned in this paper, our hierarchy could be useful for making a Sudoku puzzle with a desired level of difficulty.

We conclude the paper with the following most important property of our hierarchy. The definition of our hierarchy is purely mathematical, it does not contain any property which is described in terms of heuristic strategies such as XY-wing, XY-chain, etc. Therefor our work is different from any other existing work concerning difficulty of Sudoku puzzles such as [1]

References

1. Chang, C., Fan, Z., Sun, Y.: Hsolve: A Difficulty Metric and Puzzle Generator for Sudoku (2008), http://web.mit.edu/yisun/www/papers/sudoku.pdf
2. Gago-Vargas, J., Hartillo-Hermoso, I., Martín-Morales, J., Ucha-Enríquez, J.M.: Sudokus and Gröbner Bases: Not Only a *Divertimento*. In: Ganzha, V.G., Mayr, E.W., Vorozhtsov, E.V. (eds.) CASC 2006. LNCS, vol. 4194, pp. 155–165. Springer, Heidelberg (2006)

3. Inoue, S.: On the Computation of Comprehensive Boolean Gröbner Bases. In: Gerdt, V.P., Mayr, E.W., Vorozhtsov, E.V. (eds.) CASC 2009. LNCS, vol. 5743, pp. 130–141. Springer, Heidelberg (2009)
4. Nagai, A., Inoue, S.: An Implementation Method of Boolean Gröbner Bases and Comprehensive Boolean Gröbner Bases on General Computer Algebra Systems. In: Hong, H., Yap, C. (eds.) ICMS 2014. LNCS, vol. 8592, pp. 531–536. Springer, Heidelberg (2014)
5. Challenge your brain: is this the world's hardest Sudoku?, http://www.efamol.com/efamol-news/news-item.php?id=10
6. Introducing the World's Hardest Sudoku, http://www.efamol.com/efamol-news/news-item.php?id=43
7. Kandri-Rody, A., Kapur, D., Narendran, P.: An Ideal-Theoretic Approach for Word Problems and Unification Problems over Commutative Algebras. In: Jouannaud, J.-P. (ed.) RTA 1985. LNCS, vol. 202, pp. 345–364. Springer, Heidelberg (1985)
8. Noro, M., et al.: A Computer Algebra System Risa/Asir (2014), http://www.math.kobe-u.ac.jp/Asir/asir.html
9. Rudeanu, S.: Boolean functions and equations. North-Holland Publishing Co., American Elsevier Publishing Co., Inc., Amsterdam-London, New York (1974)
10. Sato, Y., et al.: Boolean Gröbner bases. Journal of Symbolic Computation 46(5), 622–632 (2011)
11. Sato, Y., Nagai, A., Inoue, S.: On the Computation of Elimination Ideals of Boolean Polynomial Rings. In: Kapur, D. (ed.) ASCM 2007. LNCS (LNAI), vol. 5081, pp. 334–348. Springer, Heidelberg (2008)
12. Gohnai, K.: Cross Word editorial desk (2008). Number Placement Puzzles (Basic, Middle, High, SuperHigh, Hard, SuperHard, UltraHard). Kosaido Publishing Co. (2008) (in Japanese)

A Simple GUI for Developing Applications That Use Mathematical Software Systems

Eugenio Roanes-Lozano[1] and Antonio Hernando[2]

[1] Instituto de Matemática Interdiciplinar (IMI) &
Depto. de Álgebra, Fac. de Educación, Universidad Complutense de Madrid,
c/ Rector Royo Villanova s/n, 28040–Madrid, Spain
eroanes@mat.ucm.es
http://www.ucm.es/info/secdealg/ERL/
[2] Depto. de Sistemas Inteligentes Aplicados, Escuela Universitaria de Informática,
Universidad Politécnica de Madrid,
Carretera de Valencia km 7, 28031-Madrid, Spain
ahernando@eui.upm.es

Abstract. We have observed that many mathematics teachers (of different educational levels), that know and use a certain mathematical software (like $Maple^{TM}$, *Maxima*, *MATLAB®*,...) and consider it useful for classroom use, do not use it in the classroom because they do not want to spend the time required to introduce the mathematical software system. Simultaneously, many small specific purpose mathematical applications (usually applets), that take much time to be programmed because they do not take advantage of the possibilities of existing software, have been developed. Therefore, we have implemented a humble GUI that can call different pieces of mathematical software, mainly computer algebra systems. This way, anyone with some experience programming one of the compatible pieces of mathematics software can effortlessly develop easy-to-use specific purpose applications (where the end user, for instance a student, only has to type the data or code indicated by the GUI, without having to learn in detail the mathematical software own syntax). Accessing Rule Based Expert Systems is an example of application. The GUI is compatible with most mathematical software systems that have command line versions using standard text interfaces, but requires *Windows®* to be the operating system chosen. The GUI is freely available from the authors' web page. This is a hot topic. For instance, the last release of $Maple^{TM}$, $Maple^{TM}18$, greatly increases its possibilities in this line, developing the so called "embedded components", "Clickable Math^{TM}" and "Explore" command.

Keywords: Mathematical Software Systems, Computer Algebra Systems, Graphic User Interfaces, Software Development, Rule Based Expert Systems, Mathematics Teaching.

G.A. Aranda-Corral et al. (Eds.): AISC 2014, LNAI 8884, pp. 99–119, 2014.
© Springer International Publishing Switzerland 2014

1   Introduction

1.1   Motivation: On the Need for Such a Piece of Software

We have some experience in Graphic User Interfaces (GUI) design and development. During the last fifteen years we have designed different Rule Based Expert Systems (RBES), using Gröbner Bases (GB)-based inference engines implemented on Computer Algebra Sytems (CAS) [1–3]. As most of the users of these RBES and CAS applications are neither mathematicians nor computer scientists, either $Maple^{TM}$ $Maplets^{TM}$ or different $Visual\text{-}Basic$® or $Java^{TM}$ GUI were developed ad-hoc.

Some time ago we attended a talk summarizing a research project about mathematical software development for primary schools. This team was devoted to develop $Java^{TM}$ applets for tasks that are already implemented in all CAS. They were producing very nice software, but only after a hard work. They did not directly use a friendly mathematical software systems like the (discontinued) $Derive^{TM}$, because they considered it too complex for students at the age their software was intended and because the software had to be free. We could not think of any CAS that directly matched their needs, as the free CAS we know: $Maxima$ [4], $Reduce$ [5], $CoCoA$ [6], $Axiom^{TM}$, $MuPAD$® $Light$ (also discontinued), $YACAS$ [7], $Risa/Asir$ [8], $Singular$ [9]... are not as easy to use as $Derive^{TM}$. The simplest to use is possibly $wxMaxima$ [10], which first versions looked relatively similar to $Derive^{TM}$...

We have found that many other educational authors are also developing this kind of small specific purpose tools (as may be checked searching on the web for "software" and an elementary mathematical topic).

At university level the problem is usually the lack of time. Many teachers that use mathematical software systems for their research, and consider them useful for classroom use, cannot spend the required time to introduce the mathematical software system to the students.

There are many GUI similar to the one introduced in this paper (see Section 4), but this one has the advantages of its simplicity (but the drawback of providing a fixed interface).

Obviously, if the mathematical software system chosen is not free, the user needs to have a license for that piece of software.

1.2   Background

Developing GUI for mathematical software has been an active field for a long time. A summary of Section 4 can be found immediately afterwards.

Already in 1963 Marvin Minsky's proposed program proposed a mathematical GUI. More recently we could underline $WIMS$ (WWW $Interactive$ $Multipurpose$ $Server$), a whole free system designed for developing lessons, $mathematical$ $tools$, examples... accessible via Internet.

Another powerful approach is WMI, a set of PHP scripts that interfaces with some CAS and other environment including Internet-accesible infrastructure for classroom ready material and assessment is WME.

The main approach in this line is possibly *SAGE* (*Software for Algebra and Geometry Experimentation*), a *Python*® implementation that provides an environment that can call different pieces of software from the same session. It includes the possibility to build GUIs.

This is a hot topic. For example, the last release of $Maple^{TM}$, $Maple^{TM}18$, clearly increases its possibilities in the development of specific purpose applications. For instance, the so called "Clickable math^{TM}" allows to interact with the worksheet. For example, it is possible to ask the system to solve an equation step-by-step, just by clicking on the equation (Figure 1). Another novelty are the "embedded components", that allow to easily build specific purpose GUIs executed under $Maple^{TM}18$ (Figure 2). Finally, the `Explore` command allows to easily include sliders and similar controls in specific purpose GUIs executed under $Maple^{TM}18$ (Figure 3).

1.3 Structure of the Article

The paper is structured in seven sections. After the present Introduction, the developed GUI is described in detail in Section 2. In Section 3 brief examples of the GUI calling the systems *Risa/Asir*, *Octave* [11], $Maple^{TM}$, *Maxima*, *Singular* and *CoCoA* are detailed. Section 4 gives a survey of the many existent similar works, meanwhile Section 5 underlines the main differences w.r.t. our approach. Finally, Section 6 and Section 7 contains the Conclusions and Acknowledgments, respectively.

2   The Proposed GUI

2.1 An Overview of the GUI

We believe that it would be desirable to provide of a simple GUI with:

- a text input line
- a text output window
- 2D and 3D graphic windows

i.e., a $Derive^{TM}$-style (single-line entry) GUI, that could perform any task specified (in a separate file) to the chosen CAS.

We have developed a GUI in *Visual-Basic*® for *Windows*® operating system, that can communicate with many pieces of mathematical software that have command line versions using standard text interfaces (Figure 4). The file is named GeneralGUI.EXE. It is known to work, at least, with the command line versions of $Maple^{TM}$, *Maxima*, *CoCoA*, *Risa/Asir*, *Singular* and *Octave*.

The GUI uses two files, TECHDATA.TXT and another text file (denoted, for example, CODE.TXT), which must be created by the application developer and allocated in the same directory as GeneralGUI.EXE.

TECHDATA.TXT should contain the following lines of code:

- the title of the application (to appear in the upper blue bar of the GUI),

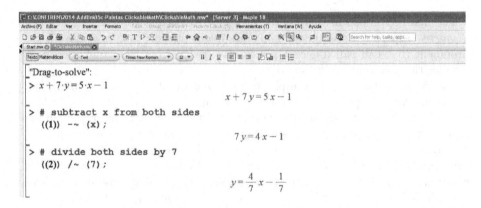

Fig. 1. An example of use of $Maple^{TM}18$'s "Clickable Math^{TM}": solving an algebraic equation step by step

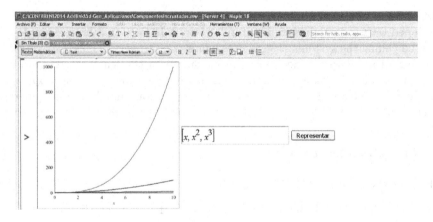

Fig. 2. A simple "embedded component" designed using $Maple^{TM}18$

- a message to the final user (to appear in the upper left corner, beside the input window),
- the path to the executable of the mathematical software system, and the name of this file, for instance:
  `C:\Program Files (x86)\Maxima-5.30.0\bin\maxima.bat`
- the prompts of the lines that should be ignored by the GUI (e.g. ">" for $Maple^{TM}$ input lines or "−−" for $CoCoA$ comments).

Meanwhile CODE.TXT contains the code to be executed by the GUI (in the syntax of the chosen piece of mathematical software).

In old *Windows*® versions (*XP* included), *Microsoft*® *.NET Framework 3.0* redistributable package [12] has to be installed. This is not required in *Vista*, *7* and *8* versions. In some computers using version *8* it will be necessary to

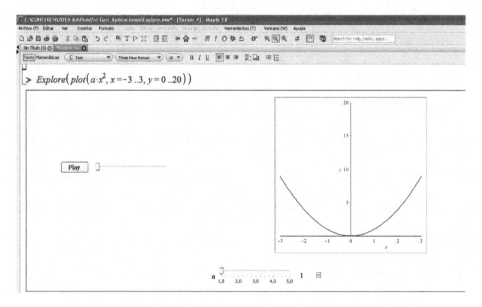

Fig. 3. A simple GUI designed using $Maple^{TM}18$'s Explore command

install the programs in the directory where the mathematical software system is installed (as happens in general with the CAS *Singular 3-x-x*; see Section 3.5).

The GUI is freely available from www.ucm.es/info/secdealg/GUI-CAS.zip (only for non-commercial use).

2.2 Plots and the GUI

There are two options if a graph is to be plotted:

- either a different window is opened from the GUI (this is the case if *Maxima* is used: it automatically calls the plotting utility *gnuplot*, that opens a new window),
- or a different version of the GUI, GeneralGUIp.EXE, that splits the space under the input line between the "algebra window" and the "plot window", is used (this is the case, for instance, if $Maple^{TM}$ is chosen: the plot is saved in a file and the GUI loads and displays it).

The complete approach is resumed in Figure 5.

2.3 Using the GUI

When the final user starts the GUI, it begins by reading TECHDATA.TXT file. It shows afterwards the main GUI window with the title and message specified by the application developer and the (empty) input and output windows within the main GUI window.

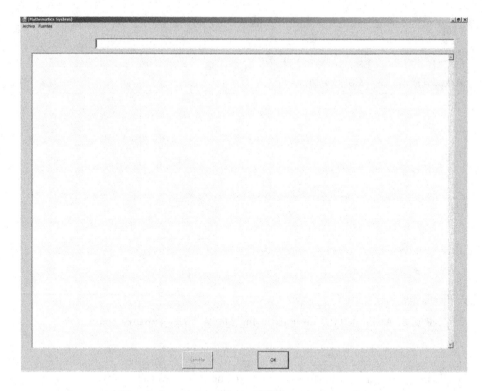

Fig. 4. Screenshot of our GUI (empty example)

When the final user types an input and clicks the "OK" or presses the Return key, a session of the chosen CAS is started and the input is stored in a global variable (VAR1). Then the code in CODE.TXT is read from the CAS session and executed for the value of VAR1.

Finally the output of the CAS is shown in the output window of the GUI. Depending on the final user's choice, the program exits or takes another input to be computed.

Note that the end user, for instance a student, only has to type the data or code indicated in the GUI, without having to learn in detail the mathematical software own syntax.

3   Examples

Let us observe that the following examples just try to show some possibilities of our GUI using different mathematical software systems in situations where the code is short, so that they can be included here. It neither tries to show all the potential of the mathematical software systems used, nor claims that the particular mathematical software system chosen in each example is the best possible one for that task.

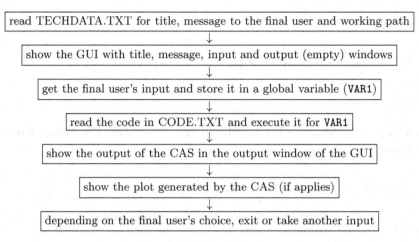

read TECHDATA.TXT for title, message to the final user and working path
↓
show the GUI with title, message, input and output (empty) windows
↓
get the final user's input and store it in a global variable (VAR1)
↓
read the code in CODE.TXT and execute it for VAR1
↓
show the output of the CAS in the output window of the GUI
↓
show the plot generated by the CAS (if applies)
↓
depending on the final user's choice, exit or take another input

Fig. 5. Algorithm of the whole process

3.1 Real Factorization of a Polynomial Using *Risa/Asir*

The application which corresponds to the GUI appearing in Figure 6 factorizes in $\mathbb{Q}$ the given polynomial (making a call to *Command Line Risa/Asir v. 20091015*, that provides the code for factorizing the polynomial).

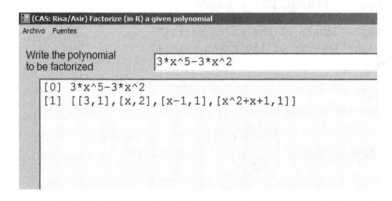

Fig. 6. Factorizing a polynomial using *Risa/Asir*. In order to make the text clearly visible, only the upper left corner of the GUI is shown.

The content of TECHDATA.TXT is, in this case:

```
(CAS: Risa/Asir) Factorize (in R) a given polynomial
Write the polynomial to be factorized
C:\Asir\Bin\asir.exe
```

```
CODE_Asir.txt
----
```

and the content of CODE.TXT is simply:

```
A=VAR1;
fctr(A);
```

In Figure 6, the final user has introduced $3x^5 - 3x^2$ and obtains its factorization as output.

3.2 Square and Inverse of a Matrix Using *Octave*

Octave is a command line system with *MATLAB®*-like syntax and possibilities.

In this case the final user introduces a square matrix and the GUI returns its square and inverse (computed numerically by *Octave 3.2.4*). Once more, as we count with the power of a mathematical software system, this is straightforward.

The content of TECHDATA.TXT is:

```
(System: Octave 3.2.4) Square and inverse of a matrix (numeric)
Write the matrix, e.g.: [[2,3];[5,1]]
C:\Octave\3.2.4_gcc-4.4.0\bin\octave-3.2.4.exe
CODE_Octave.txt
ans =
```

and the content of CODE.TXT is simply:

```
global msg1="Square of the given matrix"
global msg2="Inverse of the given matrix"
disp(msg1)
VAR1^2
disp(msg2)
VAR1^(-1)
```

The output for matrix $\begin{pmatrix} 1 & 2 & 3 \\ 1 & 1 & 1 \\ 0 & 7 & 1 \end{pmatrix}$ is shown in Figure 7.

3.3 Checking Whether an Algebraic Curve is Contained in an Algebraic Surface Using *Maple*^{TM}

Checking whether an algebraic curve is contained in an algebraic surface is straightforward using Gröbner Bases (GB) [13–15]. If the curve is the intersection of surfaces $s_1 = 0$ and $s_2 = 0$, it is clear that it is contained in the surface $s_3 = 0$ if and only if the condition $s_3 = 0$ does not add any new constraint to the algebraic system $\{s_1 = 0, s_2 = 0\}$. In terms of ideals , this is equivalent to

$$\langle s_1, s_2 \rangle = \langle s_1, s_2, s_3 \rangle$$

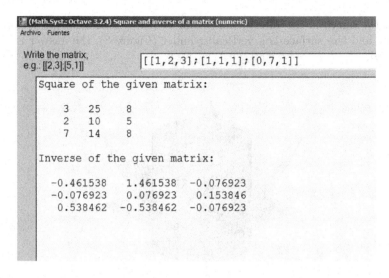

Fig. 7. Calculating the square and inverse of a matrix using *Octave*

which holds if and only if

$$GB(\langle s_1, s_2 \rangle) = GB(\langle s_1, s_2, s_3 \rangle) \ .$$

The application makes a call to *Command Line Maple*^{TM} *18* (that provides the GB code).

The content of TECHDATA.TXT is:

```
(CAS: Maple 18) Check wether an Algebraic Curve is Contained in an
                                          Algebraic Surface
Input: {eq1_curve , eq2_curve} , eq_surface
C:\Archivos de Programa\Maple 18\bin.win\cmaple.exe
CODE_Maple.txt
True
>
```

and the content of CODE.TXT is:

```
interface(warnlevel=0):
with(Groebner):
n:=VAR1:
B1:=Basis( [ op(n[1]) ] , plex(x,y,z) ):
B2:=Basis( [ op(n[1]), n[2] ] , plex(x,y,z) ):
print();
evalb(B1=B2);
```

For instance, if the final user introduces

```
{x^2+y^2-z-1,z-1},x^2+y^2-2
```

that is, the curve is the intersection of an elliptic paraboloid and a horizontal plane and the surface is a vertical cylinder (Figure 8), the answer is "Yes" (Figure 9).

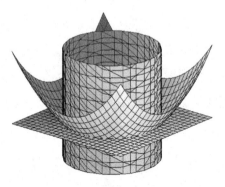

Fig. 8. The algebraic curve (given as intersection of two algebraic surfaces) and the algebraic surface

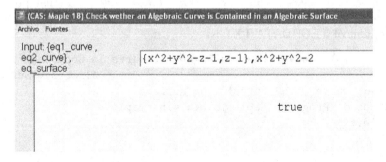

Fig. 9. Checking with $Maple^{TM}$ whether the algebraic curve is contained in the algebraic surface or not

3.4   Plotting a Function and its Derivative Using *Maxima*

This application calls *Maxima 5.30.0*, that computes a primitive of the function, displays it, and plots both the original and the primitive functions (*Maxima* calls the plotting utility *gnuplot*, that opens a new plot window, as said above).

The content of TECHDATA.TXT is, in this case:

```
(CAS: Maxima 5.30.0) Differentiate f(x) and plot f(x) and f'(x)
                                                       in (-5,5)
```

Write the function: C:\Program Files
(x86)\Maxima-5.30.0\bin\maxima.bat CODE_Maxima.txt
----

and the content of CODE.TXT is simply:

```
fun:VAR1$
string(fun);
fund:diff(fun,x)$
string(fund);
plot2d([fun,fund],[x,-5,5],[y,-5,5]);
```

that is, the function is defined, it is differentiated and both functions are plotted by *gnuplot* in a separate window (*gnuplot* should be closed before plotting another curve).

A screenshot of the output for the input $sin(x^2)$ can be found in Figure 10.

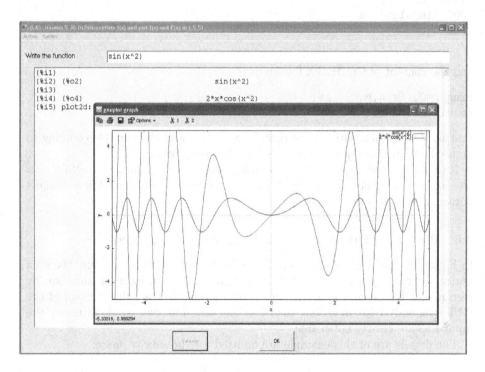

Fig. 10. Computing the derivative of a function and plotting both the function and its derivative using *Maxima* anf *gnuplot*

3.5   Compute the Resultant of Two Polynomials in x,y,z w.r.t. A Given Variable Using *Singular*

This application calls *Singular 3-1-6*, that includes a command for computing resultants.

This case is slightly more complex than the previous ones because, unlike *Singular 2-x-x*, *Singular 3-x-x* uses BASH.EXE to start *Singular* and requires:

- all files related to the GUI to be allocated where the binaries of *Cygwin* are installed, for example C:\cygwin\bin,
- to include in this directory a new BAT file, denoted, for example, Singu.BAT, containing: `bash.exe Singular`

The content of the rest of the files is similar to those above. The content of TECHDATA.TXT is:

```
(CAS: Singular 3-1-6) Resultant of two polynomials w.r.t. a
                                                     variable
Write: polynomial1, polynomial2, variable
Singu.bat
CODE_Singular.txt
True
//
```

and the content of CODE.TXT is simply:

```
ring r=0, (x,a,b,c), lp;
resultant(VAR1);
```

that is, the characteristic of the ring, its variables and the variable ordering to be used are specified and the resultant is then computed.

If $(x-a)(x-b), x-c$ and the variable x are given as input, $(a-c)(b-c) = ab - ac - bc + c^2$ is obtained (see Figure 11). The output is obtained in *Singular*'s notation: $ab - ac - bc + c2$.

3.6   Plotting a Function and its Derivative Using *Maple*^{TM}

In Figure 12 *Maple*^{TM} 18 and its package *Student* are used to compute the area under a curve, approximate a definite integral (calculate its approximation by rectangular boxes) and plot the curve and the boxes. The other version of the GUI (GeneralGUIp) has been used in this case. The two windows under the input line are clearly noticeable.

The details are of this example are omitted for the sake of space.

3.7   The First Application Developed: Consistency Checking and Knowledge Extraction in RBES

This GUI was introduced in [16] and cited in [17] as part of a shell for RBES implementation using *CoCoA 4.3* (see Figure 13).

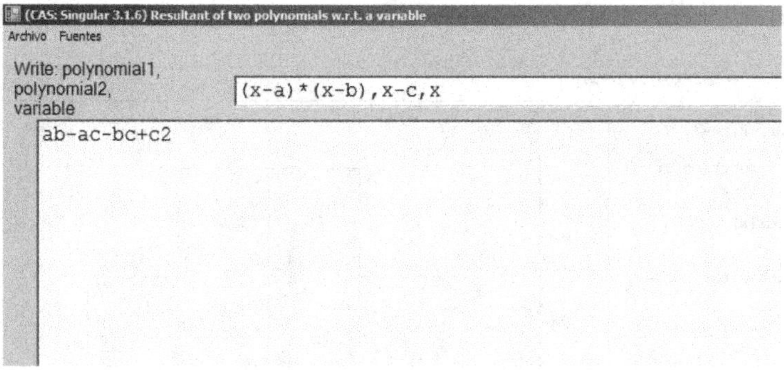

Fig. 11. Screenshot of a GUI that obtains resultants using *Singular*

Nevertheless, we believe that the GUI has a much wider range of applications, from elementary to graduate level mathematics, as shown in the other examples above.

4   Related Works

Developing GUI for available mathematical software is a very active field and many different approaches have been followed.

This section neither tries to be exhaustive nor to go into detail but to give an overview of other related works and approaches. A detailed overview of approaches and systems can be found in [18] (parts of it are obsolete; for instance, at the time, hypertextual access to documentation was only provided by a few CAS). Curiously, the importance of using *standards* like *OpenMath* for exchanging mathematical expressions was already underlined in that paper.

A more modern and exhaustive report from the same authors can be found in the first part of [19]. An overview of modern projects and systems for making mathematics available on the Internet can be found in [20].

Marvin Minsky's 1963 *Mathscope* proposed program [21] is a very early paper about a GUI for displaying publication-quality mathematical expressions and their symbolic manipulation.

In 1986 Neil M. Soiffer developed an improved GUI for the CAS *Reduce* denoted *MathScribe* [22]. It provided two-dimensional input and output of expressions in a windows environment.

In the late '80s, there was an attempt from within *Maple*^{TM} environment (*Iris*), and inspired by *MathScribe* [22], to improve access to the existing version of *Maple*^{TM} at the time [23].

In 1992 Norbert Kajler presented *CAS/PI (Computer Algebra System Portable Interface)*, which was an early attempt to provide access to different CAS (*Maple*^{TM}, *Sisyphe*, *Ulysee*) and plotting packages from a common GUI [24, 25] (it was initially developed as *Sisyphe*'s GUI). According to [24], the kernel

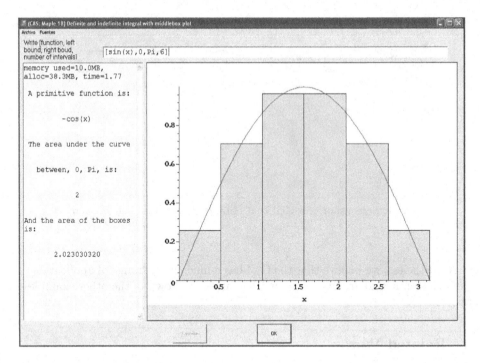

Fig. 12. A definite integral, its approximation and the corresponding plot using $Maple^{TM}$

with the formula editor requires about 17MB of memory, because the formula editor of this multi-system interface deals with syntactic differences among CAS. According to [18], "An important goal of *CAS/PI* is to allow *expert users* to tailor the user interface to specific needs and to connect it to various tools".

Some CAS were specifically designed for teaching, and subsequently included convenient GUI. The design of $Mathpert^{TM}$ [26] was based in principles like *cognitive fidelity* (the order of the steps followed by the system should be the same as those followed by the student), the *glass box principle* (the user can see how the system solves the problem) and the *correctness* principle.

Another easy-to-use system was $Theorist^{TM}$ (later renamed $LiveMath^{TM}$), that offered palettes and interactive graphics [27].

Nowadays almost all general and specific purpose CAS ($Maple^{TM}$, *Mathematica*®, *MuPAD*®, *Reduce*, *Maxima*, $Axiom^{TM}$, $Derive^{TM}$, $TI\text{-}nspire^{TM}$, *Risa/Asir*, *CoCoA*, *Singular*...) offer convenient windows-based GUI.

Nevertheless there are still attempts to improve the original GUI, like *wxMaxima* or *Kayali* [28] (these two for *Maxima*).

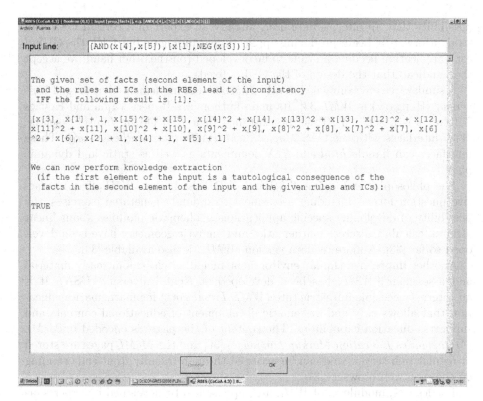

Fig. 13. Screen capture of the RBES example in [16]

Another modern attempt is *MathDrag'n* [29], a *Symbolic Equation Manipulator* that focuses on the easiness of use of the interface. It can ask *Mathematica®* or *Maxima* to perform the computations (in the background).

WIMS [30] (*WWW Interactive Multipurpose Server*) is a whole free system designed by Xiao Gang at Nice University (France) for developing lessons, *mathematical tools*, examples... accessible via Internet. The *mathematical tools*, as they are called, can communicate with some CAS (like *Maxima* or *GAP*), but the goal of the system is much broader than ours: the system includes the possibility to develop virtual classes, including mechanisms for automatic score gathering and processing. It can interface with different background programs like the CAS *Maxima*, *MuPAD®* and *GAP* as well as the mathematical software system *Octave*, the ray tracing package *POV-Ray*^{TM}, *TeX* and the proof assistant *COQ* [31]. Some *WIMS'* didactic applications are shown in [32]. There are different ongoing projects that use *WIMS* in the teaching process [30].

Focusing on *WIMS' mathematical tools*, they are far more flexible than ours (they are similar to *Maple*^{TM} *Maplets*^{TM}). Therefore, the intermediate user needs to learn how to design them, and each one takes time to be developed (see, for instance, the detailed example at the end of [32]). Whereas, we focus on the possibility that an intermediate user, already an expert in a certain CAS, can straightforwardly begin producing applications running on our GUI, the GUI of each application needing no time to be developed (on the other hand, we accept the handicap that the design of the GUI is fixed).

A similar free environment recently designed and implemented at Szeged University (Hungary) is *WMI* [33]. Its main authors are Robert Vajda and Kovács Zoltán. It is a set of PHP scripts installed in a *UNIX®* or *Linux^{TM}* server. *WMI* interfaces with the CAS *Maple^{TM}*, *MuPAD®* and *Mathematica®*. These interfaces can handle *html* and *TeX* documents as well as static and dynamic plots.

The philosophy underlying *WMI* and *WIMS* is similar: *WMI* offers interactive question forms (including assessment), randomly generated exercises, the possibility of developing specific applications... Thematic modules about linear algebra, calculus, discrete mathematics and analytic geometry have been developed so far [33]. A more modern version, *WMI2*, is also available [34].

Another impressive similar environment including classroom ready material and assessment, *WME*, has been developed at Kent University (USA). It is an Internet-accesible infrastructure (*WME Framework*) for mathematics education that allows easy and systematic development of educational contents and supports education capabilities. The content of the pages is encoded in *MeML* (*Mathematics Education Markup Language*) [36] and the *MeML* pages are stored in regular web servers and can be accessed through friendly front-ends running on common web browsers [37].

The lessons, modules... of *WME* are supposed to be developed by experts, as they "require non-trivial amounts of time, effort, and programming expertise" [35].

Also as part of the *Internet Accessible Mathematical Computation Framework* (*IAMC*) at Kent University, a page of live demos calling the CAS *Maxima* and returning *MathML* code of the output is shown [38].

Yet another big package is *SAGE* (*Software for Algebra and Geometry Experimentation*) [39–41], and its development is led by William Stein (University of Washington). It is written in *Python®* and, according to its developers, it aimed to "support research and teaching in algebra, geometry, number theory, cryptography, and related areas" and its overall goal is to "create a viable free open source alternative to *Maple^{TM}*, *Mathematica®*, *Magma* and *MATLAB®*". It provides an interface to open source mathematical software systems such as *GAP*, *Maxima*, *Singular* and *PARI/GP* and commercial mathematical software systems like *Magma*, *Mathematica®* and *Maple^{TM}*.

Probably, the key issue of *SAGE* is to provide an environment that can call different pieces of software from the same session. Therefore the user can choose the best mathematical software system for each task.

In particular, *SAGE Cell Server* [42] provides an Internet–based service for performing online computations in *SAGE*'s own language, as well as in *GAP*, *GP*, *Maxima*, *Octave*, *Python®*, *R* and *Singular*.

A system specialized in automated reasoning is *MathWeb-SB* (formerly *Math-Web*) [43–45]. It has been developed at the Universität des Saarlandes as a service infrastructure offering a uniform interface that can access different CAS (*Maple^{TM}*, *Magma*, *GAP* and *μCAS*) and automated theorem provers (*EQP*, *Otter*, *ProTein*, *Spass*, *WaldMeister*, *TPS* and *LEO*). It has been implemented in the *Mozart* programming system and uses *OpenMath* for some communication tasks. It is used in different universities in Europe and North America.

Finally, *NetSolve* [46–48] is a bridge between simple standard programming interfaces and desktop systems and the services supported by grid architecture developed at the University of Tennessee. It includes interfaces to the classic languages *C* and *Fortran^{TM}*, the mathematical software systems *MATLAB®* and *Octave* and the CAS *Mathematica®*.

With *Maple^{TM}* (*Maplets^{TM}*) or *Mathematica®* [49] it is possible to build specific purpose windows-environment (stand-alone applications cannot be created). None of the free CAS, as far as we know, offers the possibility of building specific purpose windows-environments.

In fact *Maple^{TM}* offers two further possibilities:

- *MapleNet^{TM}*, a server version of *Maple^{TM}*, that can be accessed from *Maplets^{TM}* or *Java^{TM}Applets*,
- *Maple T.A.^{TM}*, an environment for online testing and assessment

(the possibilities of both of them are summarized in the Abstract [50]). *MathML* is used to encode and display mathematical expressions. And, as said at the end of Section 1, *Maple^{TM}18* has been greatly developed its capabilities regarding application developing (see Figures 1,2,3).

Already in the late '90s *TMath* [51, 52], an extension of *Tcl* language, provided a *C++* interface to *Mathematica®* using the latter's *Mathlink®* protocol. It could also access *MATLAB®*.

JavaMath API [53, 54] is a free software that makes it possible to integrate existing CAS into Internet-accessible mathematical services. It includes interfaces to *Maple^{TM}* and *GAP* and uses the standard *OpenMath* for this interfacing. The applications need to be programmed in *Java^{TM}*.

In [55] it is described how *flash* technology can be used to develop a *middleware* for CAS. The software developed can call *Maple^{TM}* and *Mathematica®* and provides friendly interfaces (with palettes, icons...) to these CAS.

The Spanish *TutorMates®* [56, 57] includes an interface to *Maxima* and *GeoGebra*, but cannot be included in this category, as the final user cannot change or extend its content (it includes theoretical aspects, examples and exercises for Secondary Education).

5   Summary of Differences of our Approach

As we have seen above, there are different systems allowing an intermediate user to produce simple GUI for providing the final users with an easy access to a specific purpose task implemented in an existing mathematical software system.

Nevertheless, all these pieces of middleware require from the intermediate user either to master another computer language (like $Java^{TM}$) or to learn how to design the GUI in the system's own grammar.

Instead, our approach exclusively requires from the intermediate user to know to program in the mathematical software system (accepting, as a disadvantage, that the appearance of the GUI is fixed).

6   Conclusions

Summarizing, we believe that in science, engineering and education, there is a need for producing small very specific purpose applications requiring of algorithms that are already implemented in all mathematical software systems. We have developed a very easy-to-use tool which may be very convenient for mathematical software developers. The main interest of the implementation lies in software reuse.

The future development could consist in:

- adapting this general purpose mathematical GUI to other CAS
- analyzing the possibilities of migrating to other operating systems like X $Window$ $System^{TM}$ (e.g. implementing the GUI in a portable language like $Java^{TM}$)
- studying the possibility to offer the GUI as a web service,
- spreading this software to potential application developers (trying to get their feedback for future refinements).

This piece of software can be freely downloaded from

www.ucm.es/info/secdealg/GUI-CAS.zip

(only for non-commercial use).

Acknowledgements. This work was partially supported by the research project *TIN2012-32482* (Spanish Government).

The authors would also like to thank the anonymous reviewers for their valuable comments, which helped to improve the manuscript.

References

1. Pérez-Carretero, C., Laita, L.M., Roanes-Lozano, E., Lázaro, L., González-Cajal, J., Laita, L.: A Logic and Computer Algebra-Based Expert System for Diagnosis of Anorexia. Math. Comp. Simul. 58(3), 183–202 (2002)
2. Roanes-Lozano, E., López-Vidriero Jr., E., Laita, L.M., López-Vidriero, E., Maojo, V., Roanes-Macías, E.: An Expert System on Detection, Evaluation and Treatment of Hypertension. In: Buchberger, B., Campbell, J. (eds.) AISC 2004. LNCS (LNAI), vol. 3249, pp. 251–264. Springer, Heidelberg (2004)

3. Rodríguez-Solano, C., Laita, L.M., Roanes-Lozano, E., López Corral, L., Laita, L.: A Computational System for Diagnosis of Depressive Situations. Exp. Sys. Appl. 31, 47–55 (2006)
4. Maxima: Maxima, a Computer Algebra System. Version 5.30.0 (2013), http://maxima.sourceforge.net/
5. http://www.reduce-algebra.com
6. Abbot, J., Bigatti, A.M., Lagorio, G.: CoCoA-5: A system for doing Computations in Commutative Algebra, http://cocoa.dima.unige.it
7. http://yacas.soureceforge.net/homepage.html
8. http://www.math.kobe-u.ac.jp/Asir/
9. Decker, W., Greuel, G.-M., Pfister, G., Schönemann, H.: Singular 3-1-6 — A computer algebra system for polynomial computations (2012), http://www.singular.uni-kl.de
10. http://wxmaxima.sourceforge.net/
11. Octave community: GNU Octave 3.8.1 (2014), http://www.gnu.org/software/octave/
12. http://www.microsoft.com/downloads/
13. Buchberger, B.: Bruno Buchberger's PhD thesis 1965: An algorithm for finding the basis elements of the residue class ring of a zero dimensional polynomial ideal. Journal of Symbolic Computation 41(3-4), 475–511 (2006)
14. Roanes-Lozano, E., Roanes-Macías, E., Laita, L.M.: The Geometry of Algebraic Systems and Their Exact Solving Using Groebner Bases. Comp. Sci. Eng. 6(2), 76–79 (2004)
15. Roanes-Lozano, E., Roanes-Macías, E., Laita, L.M.: Some Applications of Gröbner Bases. Comp. Sci. Eng. 6(3), 56–60 (2004)
16. Roanes-Lozano, E., Hernando, A., Laita, L.M., Roanes-Macías, E.: A Shell for Rule-Based Expert Systems Development Using Groebner Bases-Based Inference Engines. In: Ruan, D., Montero, J., Lu, J., Martínez, L., D'hondt, P., Kerre, E.E. (eds.) Computational Intelligence in Decision and Control. Proceedings of the 8th International FLINS Conference. Procs. Series on Computer Engineering and Information Science, vol. 1, pp. 769–774. World Scientific, Singapore (2008)
17. Roanes-Lozano, E., Laita, L.M., Hernando, A., Roanes-Macías, E.: An algebraic approach to rule based expert systems. RACSAM 104(1), 19–40 (2010), doi:10.5052/RACSAM
18. Kajler, N., Soiffer, N.: Some Human Interaction Issues in Computer Algebra. CAN Nieuwsbrief 12, 14–24 (1994); Also in: ACM SIGSAM Bull. 28(1), 18–28 (1994), Also in: SIGCHI Bull. 26(4), 64–69 (1994)
19. Kajler, N., Soiffer, N.: A Survey of User Interfaces for Computer Algebra Systems. J. Symb. Comp. 25(2), 127–160 (1998)
20. http://wme.cs.kent.edu/research.html
21. Minsky, M.L.: MATHSCOPE: Part I - A Proposal for a Mathematical Manipulation-Display System. Technical Report MAC-M-118, Artificial Intelligence Project, Memo 61. MIT, Cambridge, MA, USA (1963)
22. Soiffer, N.M.: The Design of a User Interface for Computer Algebra Systems (Ph.D. Thesis). Report UCB/CSD/91/626 (University of California, Berkeley, CA, USA) (1991), http://www.eecs.berkeley.edu/Pubs/TechRpts/1991/CSD-91-626.pdf
23. Leong, B.L.: Iris: Design of a User Interface Program for Symbolic Algebra, in Symposium on Symbolic and Algebraic Manipulation. In: Proceedings of the Fifth ACM Symposium on Symbolic and Algebraic Computation, pp. 1–6. ACM Press, New York (1986)

24. Kajler, N.: CAS/PI: a Portable and Extensible Interface for Computer Algebra Systems. In: Proceedings of ISSAC 1992, pp. 376–386. ACM Press, New York (1992)

25. Kajler, N.: User Interfaces for Symbolic Computation: A Case Study. In: UIST 1993: Proceedings of the 6th Annual ACM Symposium on User Interface Software and Technology, pp. 1–10. ACM Press, New York (1993)

26. Beeson, M.: Design Principles of *Mathpert*: Software to Support Education in Algebra and Calculus. In: Kajler, N. (ed.) Computer-Human Interaction in Symbolic Computation, pp. 89–115. Springer, Vienna (1998)

27. Greenman, J.: Theorist: A Review. Maths&Stats 5(1), 15–17 (1994)

28. http://kayali.sourceforge.net/

29. http://sourceforge.net/projects/equation/

30. http://wims.unice.fr

31. Gang, X.: WIMS: An Interactive Mathematics Server. J. Online Math. Appl. Section 3.1 Algebra (2001),
http://www.maa.org/publications/periodicals/loci/joma/wims-an-interactive-mathematics-server

32. Galligo, A., Xiao, G.: Using WIMS for Mathematical Education. In: 2001 IAMC Workshop Proceedings (electronic), a Workshop at ISSAC 2001 (2001),
http://icm.mcs.kent.edu/research/iamc.html

33. Vajda, R., Kovacs, Z.: Interactive Web Portals in Mathematics. Teaching Math. and Comp. Sci. 1(2), 347–361 (2003)

34. http://matek.hu

35. http://wme.cs.kent.edu/

36. Wang, P.S., Zhou, Y., Zou, X.: Web-based Mathematics Education: MeML Design and Implementation. In: Proceedings of IEEE/ITCC 2004, Las Vegas, Nevada, USA, pp. 169–175 (2004)

37. Wang, P.S., Kajler, N., Zhou, Y., Zou, X.: WME: Towards a Web for Mathematics Education. In: ISSAC 2003 Proceedings, pp. 258–265. ACM Press, New York (2003)

38. http://icm.mcs.kent.edu/research/demo.html

39. Stein, W.A.: SAGE: Software for Algebra adn Geometry Experimentation. Project Description (2006),
http://modular.math.washington.edu/grants/sage-06/project_description.pdf

40. http://www.sagemath.org/

41. Stein, W., Joyner, D.: SAGE: System for Algebra and Geometry Experimentation. SISGSAM Bull. 39(2), 61–64 (2005)

42. https://sagecell.sagemath.org/

43. Franke, A., Kohlhase, M.: System Description: MathWeb, an Agent-Based Communications Layer for Distributed Automated Theorem Proving. In: Ganzinger, H. (ed.) CADE-16. LNCS (LNAI), vol. 1632, pp. 217–221. Springer, Heidelberg (1999)

44. Zimmer, J., Kohlhase, M.: System Description: The MathWeb Software Bus for Distributed Mathematical Reasoning. In: Voronkov, A. (ed.) CADE-18. LNCS (LNAI), vol. 2392, pp. 139–143. Springer, Heidelberg (2002)

45. Franke, A., Hess, S.M., Jung, C.G., Kohlhase, M., Sorge, V.: Agent-Oriented Integration of Distributed Mathematical Services. J. of Universal Comp. Sci. 5(3), 156–187 (1999)

46. http://icl.cs.utk.edu/netsolve/

47. Seymour, K., Yarkhan, A., Agrawal, S., Dongarra, J.: NetSolve: Grid Enabling Scientific Computing Environments. In: Grandinetti, L. (ed.) Grid Computing and New Frontiers of High Performance Computing. Advances in Parallel Computing Series, vol. 14, pp. 33–52. Elsevier Science, Amsterdam (2005)

48. Agrawal, S., Dongarra, J., Seymour, K., Vadhiyar, S.: NetSolve: Past, Present, and Future - A Look at a Grid Enabled Server. In: Berman, F., Fox, G., Hey, A. (eds.) Grid Computing - Making the Global Infrastructure a Reality, pp. 613–622. John Wiley & Sons Ltd., Chichester (2003)

49. Anonymous: GUIKit, Technical software news (Wolfram Research) 2, 2 (2004)

50. Bernardin, L.: Mathematical Computations on the Web: The Maple Approach. In: 2004 IAMC Workshop Proceedings (electronic), a Workshop at ISSAC 2004 (2001), http://www.orcca.on.ca/conferences/iamc2004/abstracts/index.html

51. http://library.wolfram.com/infocenter/MathSource/642

52. http://ptolemy.eecs.berkeley.edu/other/tmath/tmath0.2/README.html

53. http://javamath.sourceforge.net/

54. Solomon, A., Struble, C.A., Cooper, A., Linton, S.A.: The JavaMath API: An architecture for Internet accessible mathematical services. Submitted to the J. Symb. Comp.

55. Song, K.: Flash-Enabled User Interface for CAS. In: 2003 IAMC Workshop Proceedings (electronic), a Workshop at ISSAC 2003 (2003), http://www.researchgate.net/publication/240915416_Flash-Enabled_User_Interface_For_CAS

56. http://www.tutormates.es

57. Grupo TutorMates: TutorMates: Cálculo Científico en Educación Secundaria. La Gaceta de la RSME 14(3), 565–577 (2011)

Conformant Planning as a Case Study
of Incremental QBF Solving[*]

Uwe Egly[1], Martin Kronegger[1], Florian Lonsing[1], and Andreas Pfandler[1,2]

[1] Institute of Information Systems, Vienna University of Technology, Austria
[2] School of Economic Disciplines, University of Siegen, Germany
{firstname.lastname}@tuwien.ac.at

Abstract. We consider planning with uncertainty in the initial state as a case study of *incremental* quantified Boolean formula (QBF) solving. We report on experiments with a workflow to incrementally encode a planning instance into a sequence of QBFs. To solve this sequence of successively constructed QBFs, we use our *general-purpose incremental* QBF solver DepQBF. Since the generated QBFs have many clauses and variables in common, our approach avoids redundancy both in the encoding phase and in the solving phase. Experimental results show that incremental QBF solving outperforms non-incremental QBF solving. Our results are the first empirical study of *incremental* QBF solving in the context of planning and motivate its use in other application domains.

1   Introduction

Many workflows in formal verification and model checking rely on certain logics as languages to model verification conditions or properties of the systems under consideration. Examples are propositional logic (SAT), quantified Boolean formulas (QBFs), and decidable fragments of first order logic in terms of satisfiability modulo theories (SMT). A tight integration of decision procedures to solve formulas in these logics is crucial for the overall performance of the workflows in practice.

In the context of SAT, *incremental solving* has become a state of the art approach [1,10,20,28]. Given a sequence of related propositional formulas $S = \langle \phi_0, \phi_1, \ldots, \phi_n \rangle$ an incremental SAT solver reuses information that was gathered when solving ϕ_i in order to solve the next formula ϕ_{i+1}. Since incremental solving avoids some redundancy in the process of solving the sequence S, it is desirable to integrate incremental solvers in practical workflows. In contrast, in *non-incremental solving* the solver does not keep any information from previously solved formulas and always starts from scratch.

QBFs allow for explicit universal ($\forall$) and existential ($\exists$) quantification over Boolean variables. The problem of checking the satisfiability of QBFs is PSPACE-complete. We consider QBFs as a natural modelling language for planning problems with uncertainty in the initial state. In *conformant planning* we are given a set of state variables over a specified domain, a set of actions with preconditions and effects, an initial state where some values of the variables may be unknown, and a specification of the goal.

[*] Supported by the Austrian Science Fund (FWF) under grants S11409-N23 and P25518-N23 and the German Research Foundation (DFG) under grant ER 738/2-1.

G.A. Aranda-Corral et al. (Eds.): AISC 2014, LNAI 8884, pp. 120–131, 2014.
© Springer International Publishing Switzerland 2014

The task is to find a sequence of actions, i.e., a plan, that leads from the initial state to a state where the goal is satisfied. Many natural problems, such as repair and therapy planning [33], can be formulated as conformant planning problems. When restricted to plans of length polynomial in the input size this form of planning is $\Sigma_2 P$-complete [3], whereas classical planning is NP-complete.

Therefore, using a transformation to QBFs in the case of conformant planning is a very natural approach. Rintanen [31] presented such transformations. Recently, Kronegger et al. [18] showed that transforming the planning instance into a sequence of QBFs can be competitive. In this approach, they generated a QBF for every plan length under consideration and invoked an external QBF solver on each generated QBF. However, the major drawback is that the QBF solver cannot reuse information from previous runs and thus has to relearn all necessary information in order to solve the QBF. In this work we overcome this problem by tightly integrating a *general-purpose incremental QBF solver* in an incremental workflow to solve planning problems. To obtain a better picture of the performance gain through the incremental approach, we perform a case study where we compare incremental and non-incremental QBF solving on benchmarks for conformant planning.

The **main contributions** of this work are as follows.

- *Planning tool.* We present a planning tool based on the transformation of planning instances with unknown variables in the initial state to QBFs. This tool implements an incremental and exact approach, i.e., it is guaranteed to find a plan whenever a plan exists and – if successful – it returns a plan of *minimal* length. Furthermore, our tool allows for the use of arbitrary (incremental) QBF solvers.
- *Experimental evaluation.* We evaluate the performance of the incremental and the non-incremental approach to planning with incomplete information in the initial state. Thereby, we rely on incremental and non-incremental variants of the QBF solver DepQBF [23,24].[1] Incremental QBF solving outperforms non-incremental QBF solving in our planning tool. Our results are a case study of incremental QBF solving and motivate its use in other application domains. In addition, we also compare our results to heuristic approaches.

2 Incremental QBF Solving

We focus on QBFs $\psi = \hat{Q}.\phi$ in prenex conjunctive normal form (PCNF). All quantifiers occur in the prefix $\hat{Q} = Q_1 B_1 \ldots Q_n B_n$ and the CNF part ϕ is a quantifier-free propositional formula in CNF. The prefix consists of pairwise disjoint sets B_i of quantified Boolean variables, where $Q_i \in \{\forall, \exists\}$, and gives rise to a linear ordering of the variables: we define $x < y$ if $x \in B_i, y \in B_j$ and $i < j$.

The semantics of QBFs is defined recursively based on the quantifier types and the prefix ordering of the variables. The QBF consisting only of the truth constant *true* ($\top$) or *false* ($\bot$) is satisfiable or unsatisfiable, respectively. The QBF $\psi = \forall x. \psi'$ with the universal quantification $\forall x$ at the leftmost position in the prefix is satisfiable if $\psi[x := \bot]$ and $\psi[x := \top]$ are satisfiable, where the formula $\psi[x := \bot]$ ($\psi[x := \top]$)

---

[1] DepQBF is free software: http://lonsing.github.io/depqbf/

results from ψ by replacing the free variable x by $\bot$ ($\top$). The QBF $\psi = \exists x.\, \psi'$ with the existential quantification $\exists x$ is satisfiable if $\psi[x := \bot]$ or $\psi[x := \top]$ are satisfiable.

Search-based QBF solving [8] is a generalization of the DPLL algorithm [9] for SAT. Modern search-based QBF solvers implement a QBF-specific variant of conflict-driven clause learning (CDCL) for SAT, called QCDCL [11,21,25,34]. In QCDCL the variables are successively assigned until an (un)satisfiable subcase is encountered. The subcase is analyzed and new learned constraints (clauses or cubes) are inferred by Q-resolution [17,25]. The purpose of the learned constraints is to prune the search space and to speed up proof search. Assignments are retracted by backtracking and the next subcase is determined until the formula is solved.

Let $\langle \psi_0, \psi_1, \ldots, \psi_n \rangle$ be a sequence of QBFs. In *incremental QBF solving based on QCDCL*, we must keep track which of the constraints that were learned on a solved QBF ψ_i can be reused for solving the QBFs ψ_j with $i < j$. An approach to incremental QBF solving was first presented in the context of bounded model checking [27]. We rely on the general-purpose incremental QBF solver DepQBF [23,24].

To illustrate the potential of incremental QBF solving, we present a case study of QBF-based conformant planning in the following sections. To this end we discuss conformant planning and two types of benchmarks used in the experimental analysis.

3   Conformant Planning and Benchmark Domains

A conformant planning problem consists of a set of state variables over a specified domain, a set of actions with preconditions and effects, an initial state where some values of the variables may be unknown, and a specification of the goal. The task is to find a sequence of actions, i.e., a plan, that leads from the initial state to a state where the goal is satisfied. The plan has to reach the goal for all possible values of unknown variables, i.e., it has to be fail-safe. This problem can nicely be encoded into QBFs, e.g., by building upon the encodings by Rintanen [31]. Conformant planning naturally arises, e.g., in repair and therapy planning [33], where a plan needs to succeed even if some obstacles arise.

The length of a plan is the number of actions in the plan. As one is usually looking for short plans, the following strategy is used. Starting at a lower bound k on the minimal plan length, we iteratively increment the plan length k until a plan is found or a limit on the plan length is reached. This strategy is readily supported by an incremental QBF solver because a large number of clauses remains untouched when moving from length k to $k + 1$ and always leads to *optimal* plans with respect to the plan length.

The two benchmark types we consider in our case study are called "Dungeon". These benchmarks are inspired by adventure computer-games and were first presented at the QBF workshop 2013 [18]. In this setting a player wants to defeat monsters living in a dungeon. Each monster requires a certain configuration of items to be defeated. In the beginning, the player picks at most one item from each pool of items. In addition, the player can exchange several items for one more powerful item if she holds all necessary "ingredients". Eventually, the player enters the dungeon. When entering the dungeon, the player is forced to pick additional items. The dilemma is that the player does not know which items she will get, i.e., the additional items are represented by variables

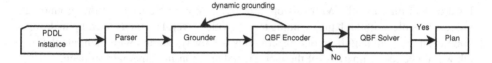

dynamic grounding

Fig. 1. Architecture of the planning tool

with *unknown* values in the initial state. It might also happen that the new items turn out to be obstructive given the previously chosen item configuration. The goal is to pick items such that irrespective of the additional items she defeats at least one monster.

We consider two variants of the Dungeon benchmark. In variant *v0* the player is only allowed to enter the dungeon once, thus has to pick the items and build more powerful items in advance. In contrast, in variant *v1* the player might attempt fighting the monsters several times and pick/build further items in between if she was unsuccessful.

Despite the simple concept, these benchmarks are well suited for our case study. First, they capture the full hardness of $\Sigma_2 P$-complete problems. Second, it is natural to reinterpret the game setting as a configuration or maintenance problem.

4 QBF Planning Tool

We briefly describe our planning tool that takes planning instances as input and encodes them as a sequence of QBFs. This tool generates a plan of minimal length for a given conformant planning instance with uncertainty in the initial state.

Figure 1 illustrates the architecture of our planning tool which was used for the experiments. The tool takes a planning instances given in PDDL format as input. After parsing the input, the grounder analyzes the given planning instance and calculates a lower bound ℓ on the plan length. Starting with a plan length of $k = \ell$, the grounder then grounds only relevant parts of the instance, i.e., the grounder systematically eliminates variables from the PDDL instance. In a next step, the QBF encoder takes the ground representation as input and transforms it into a QBF that is satisfiable if and only if the planning problem has a plan of length k. The encoding which is used for this transformation to QBFs builds upon the ∃∀∃-encoding described in the work of Rintanen [31]. We decided to employ the ∃∀∃-encoding rather than a ∃∀-encoding as this gives a more natural encoding and simplifies the PCNF transformation. Since in this work we focus on a comparison of the incremental and non-incremental approach, we do not go into the details of the encoding. After the transformation into a QBF, the QBF encoder then invokes a QBF solver on the generated QBF. If the generated QBF is satisfiable, our system extracts the optimal plan from the assignment of the leftmost ∃-block. If the QBF is unsatisfiable, the plan length k is incremented, additional relevant parts of the problem may need grounding, and the subsequent QBF is passed to the solver. Below, we give an overview of the features and optimizations of our planning tool.

Since grounding the planning instance can cause an exponential blow-up in the size of the input, we have implemented a dynamic grounding algorithm. This algorithm uses ideas from the concept of the planning graph [6] to only ground actions that are relevant for a certain plan length. With this optimization, we are able to make the grounding process feasible. Although the planning tool provides several methods to compute lower bounds on the plan length, in our experiments we always started with plan length 0 to allow for a better comparison of the incremental and non-incremental approach.

Our *incremental* QBF solver DepQBF is written in C whereas the planning tool is written in Java. To integrate DepQBF in our tool and to employ its features for incremental solving, we implemented a Java interface for DepQBF, called *DepQBF4J*.[2] This way, DepQBF can be integrated into arbitrary Java applications and its API functions can then be called via the Java Native Interface (JNI).

In our planning tool, the use of DepQBF's API is crucial for incremental solving because we have to avoid writing the generated QBFs to a file. Instead, we add and modify the QBFs to be solved directly via the API of DepQBF. The API provides *push* and *pop* functions to add and remove *frames*, i.e., sets of clauses, in a stack-based manner. The CNF part of a QBF is represented as a sequence of frames.

Given a planning instance, the workflow starts with plan length $k = 0$. The QBF ψ_k for plan length k can be encoded naturally in an incremental fashion by maintaining two frames f_0 and f_1 of clauses: clauses which encode the goal state are added to f_1. All other clauses are added to f_0. Frame f_0 is added to DepQBF before frame f_1. If ψ_k is unsatisfiable, then f_1 is deleted by a *pop* operation, i.e., the clauses encoding the goal state of plan length k are removed. The plan length is increased by one and additional clauses encoding the possible state transitions from plan length k to $k + 1$ are added to f_0. The clauses encoding the goal state for plan length $k + 1$ are added to a new f_1. Note that in the workflow clauses are added to f_0 but this frame is never deleted.

The workflow terminates if (1) the QBF ψ_k is satisfiable, indicating that the instance has a plan with *optimal* length k, or (2) ψ_k is unsatisfiable and $k + 1$ exceeds a user-defined upper bound, indicating that the instance does not have a plan of length k or smaller, or (3) the time or memory limits are exceeded. In the cases (1) and (2), we consider the planning instance as solved. For the experimental evaluation, we imposed an upper bound of 200 on the plan length.

The Dungeon benchmark captures the full hardness of problems on the second level of the polynomial hierarchy. Therefore, as shown in the following section, already instances with moderate plan lengths might be hard for QBF solvers as well as for planning-specific solvers [18]. We considered an upper bound of 200 of the plan length to be sufficient to show the difference between the incremental and non-incremental QBF-based approach. The hardness is due to the highly combinatorial nature of the Dungeon instances, which also applies to configuration and maintenance problems. Further, configuration and maintenance problems can be encoded easily into conformant planning as the Dungeon benchmark is essentially a configuration problem.

Our planning tool can also be combined with any *non-incremental* QBF solver to determine a plan of minimal length in a *non-incremental* fashion. This is done by writing the QBFs which correspond to the plan lengths $k = 0, 1, \ldots$ under consideration to separate files and solving them with a standalone QBF solver [18].

---

[2] DepQBF4J is part of the release of DepQBF version 3.03 or later.

Table 1. Overall statistics for the planning workflows implementing incremental and non-incremental QBF solving by incDepQBF and DepQBF, respectively: total time for the workflow on all 288 instances (including time outs), solved instances, solved instances where a plan was found and where no plan with length 200 or shorter exists, average time ($\bar{t}$) in seconds, number of backtracks ($\bar{b}$) and assignments ($\bar{a}$) performed by the QBF solver on the solved instances.

		288 Planning Instances (Dungeon Benchmark: v0 and v1)					
	Time	Solved	Plan found	No plan	$\bar{t}$	$\bar{b}$	$\bar{a}$
DepQBF:	112,117	168	163	5	24.40	2210	501,706
incDepQBF:	103,378	176	163	13	14.55	965	120,166

Table 2. Statistics like in Table 1 but for those planning instances which were uniquely solved when using either incDepQBF or DepQBF, respectively. For all of these uniquely solved instances, no plan was found within the given upper bound of 200.

	Uniquely Solved Planning Instances					
	Solved	Plan found	No plan	$\bar{t}$	$\bar{b}$	$\bar{a}$
DepQBF:	2	0	2	545.04	99	1,024,200
incDepQBF:	10	0	10	94.15	174	45,180

5 Experimental Evaluation

We evaluate the incremental workflow described in the previous section using planning instances from the Dungeon benchmark. The purpose of our experimental analysis is to compare incremental and non-incremental QBF solving in the context of conformant planning. Thereby, we provide the first empirical study of *incremental* QBF solving in the planning domain. In addition to [26,27], our results independently motivate the use of incremental QBF solving in other application domains.

From the Dungeon benchmark described in Section 3, we selected 144 planning instances from each variant *v0* and *v1*, resulting in 288 planning instances. Given a planning instance, we allowed 900 seconds wall clock time and 7 GB of memory for the entire workflow, which includes grounding, QBF encoding and QBF solving. All experiments reported were run on AMD Opteron 6238, 2.6 GHz, 64-bit Linux.

We first compare the performance of incremental and non-incremental QBF solving in the planning workflow. To this end, we used incremental and non-incremental variants of our QBF solver DepQBF, referred to as incDepQBF and DepQBF, respectively. For non-incremental solving, we called the standalone solver DepQBF by system calls from our planning tool. Thereby, we generated the QBF encoding of a particular planning instance and wrote it to a file on the hard disk. DepQBF then reads the QBF from the file. For incremental solving, we called incDepQBF through its API via the DepQBF4J interface. This way, the QBF encoding is directly added to incDepQBF by its API within the planning tool (as outlined in the previous section), and no files are written. The solvers incDepQBF and DepQBF have the *same* codebase. Therefore, differences in their performance are due to whether incremental solving is applied or not.

The statistics in Tables 1 to 3 and Figure 3 illustrate that incremental QBF solving by incDepQBF outperforms non-incremental solving by DepQBF in the workflow in terms

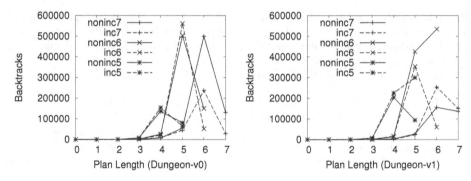

Fig. 2. Related to Table 3. Let P_5, P_6 and P_7 be the sets of planning instances where a plan of length 5, 6, and 7 was found using both incDepQBF and DepQBF. The data points on the lines "inc5" (dashed) and "noninc5" (solid) show the total numbers of backtracks spent by incDepQBF and DepQBF on the QBFs corresponding to the plan lengths $i = 0, \ldots, 5$ for all instances in P_5. The data points for P_6 and P_7 were computed similarly for the plan lengths $i = 0, \ldots, 6$ and $i = 0, \ldots, 7$, respectively, and are shown on the lines "inc6", "noninc6" and "inc7", "noninc7".

of solved instances, uniquely solved instances (Table 2), run time and in the number of backtracks and assignments spent in QBF solving. With incDepQBF and DepQBF, 166 instances were solved by both. For three of these 166 instances, no plan exists.

The different calling principles of incDepQBF (by the API) and DepQBF (by system calls) may have some influence on the overall run time of the workflow, depending on the underlying hardware and operating system. In general, the use of the API avoids I/O overhead in terms of hard disk accesses and thus might save run time. Due to the timeout of 900 seconds and the relatively small number of QBF solver calls in the workflow (at most 201, for plan length 0 up to the upper bound of 200), we expect that the influence of the calling principle on the overall time statistics in Tables 1 and 2 and Figure 3 is only marginal. Moreover, considering backtracks and assignments as shown in Table 3 as an independent measure of the performance of the workflow, incremental solving by incDepQBF clearly outperforms non-incremental solving by DepQBF.

Figure 2 shows how the number of backtracks evolves if the plan length is increased. On the selected instances which have a plan with optimal length k, we observed peaks in the number of backtracks by incDepQBF and DepQBF on those QBFs which correspond to the plan length $k - 1$. Thus empirically the final unsatisfiable QBF for plan length $k - 1$ is harder to solve than the QBF for the optimal plan length k or shorter plan lengths. Figure 2 (right) shows notable exceptions. For $k = 6$, the number of backtracks by DepQBF increases in contrast to incDepQBF. For $k = 5$ and $k = 7$, incDepQBF spent more backtracks than DepQBF. We attribute this difference to the heuristics in (inc)DepQBF. The same QBFs must be solved by incDepQBF and De-pQBF in one run of the workflow. However, the heuristics in incDepQBF might be negatively influenced by previously solved QBFs. We made similar observations on instances not solved with either incDepQBF or DepQBF where DepQBF reached a longer plan length than incDepQBF within the time limit.

Incremental solving performs particularly well on instances for which no plan exists. Considering the ten instances uniquely solved with incDepQBF (Table 2), on average it

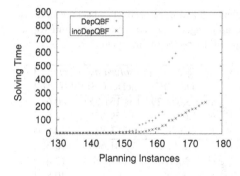 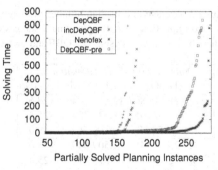

Fig. 3. Sorted run times of the workflow with incremental (incDepQBF) and non-incremental solving (DepQBF), related to Table 1

Fig. 4. Sorted accumulated run times of solvers on selected QBFs from the planning workflow. DepQBF-pre includes preprocessing.

took less than 0.5 seconds to encode and solve one of the 201 unsatisfiable QBFs (i.e., from plan length zero to the upper bound of 200) in the planning workflow. Considering the 13 instances solved using incDepQBF which do not have a plan (Table 1), on average the workflow took 73.80 seconds and incDepQBF spent 35,729 assignments and 135 backtracks. In contrast to that, the workflow using DepQBF took 270.92 seconds on average to solve the five instances which do not have a plan (Table 1), and DepQBF spent 421,619 assignments and 99 backtracks.

5.1 Preprocessing

The implementation of (inc)DepQBF does not include preprocessing [5,12]. In general, preprocessing might be very beneficial for the performance of QBF-based workflows. The efficient combination of preprocessing and incremental solving is part of ongoing research in SAT [19,28] and QBF [14,16,26,32].

In order to evaluate the *potential* impact of preprocessing in our workflow, we carried out the following experiment. We ran the workflow using DepQBF on all 288 planning instances with a time limit of 900 seconds and collected all QBFs that were generated this way. Like for the results in Table 1, we ran DepQBF and incDepQBF on these QBFs within our workflow. Additionally, we ran the QBF solver Nenofex [22] because it performed well on QBFs generated from the Dungeon benchmark.[3] Nenofex successively eliminates variables in a QBF by expansion at the cost of a possibly exponential blow up of the formula size. Figure 4 shows the run times of DepQBF, incDepQBF, Nenofex and DepQBF-pre, which combines DepQBF with the preprocessor Bloqqer [5]. We accumulated the solving times spent on QBFs that were generated from a particular planning instance. The plot shows these accumulated times for each planning instance. Run times smaller than the time out of 900 seconds do not necessarily indicate that the planning instance was solved because we considered only a subset of the QBFs corresponding

---

[3] Results of Nenofex in the QBF Gallery 2013: http://www.kr.tuwien.ac.at/events/qbfgallery2013/sc_apps/conf_planning_dungeon.html

Table 3. Average and median number of assignments ($\bar{a}$ and $\tilde{a}$, respectively), backtracks ($\bar{b}, \tilde{b}$), and workflow time ($\bar{t}, \tilde{t}$) for planning instances from Dungeon-*v0* (left) and Dungeon-*v1* (right) where both workflows using DepQBF and incDepQBF found the optimal plan

		Dungeon-v0 (81 solved instances)					Dungeon-v1 (82 solved instances)		
		DepQBF	incDepQBF	diff. (%)			DepQBF	incDepQBF	diff. (%)
Total	a:	171,245,867	122,233,046	-28.6	*Total*	a:	183,674,291	164,131,257	-10.6
	b:	1,660,296	1,237,384	-25.4		b:	1,670,375	1,459,655	-12.6
	t:	1,253.50	638.94	-49.0		t:	1,308.26	773.39	-40.8
Per instance	$\bar{a}$:	2,114,146	1,509,049	-28.6	*Per instance*	$\bar{a}$:	2,239,930	2,001,600	-10.6
	$\bar{b}$:	20,497	15,276	-25.4		$\bar{b}$:	20,370	17,800	-12.6
	$\bar{t}$:	15.47	7.88	-49.0		$\bar{t}$:	15.95	9.43	-40.8
	$\tilde{a}$:	1,388	1,391	+0.2		$\tilde{a}$:	1,595	1,641	+2.8
	$\tilde{b}$:	13	11	-15.3		$\tilde{b}$:	15	15	+0.0
	$\tilde{t}$:	1.01	0.37	-63.8		$\tilde{t}$:	1.31	0.37	-71.7
Per solved QBF	$\bar{a}$:	629,580	449,386	-28.6	*Per solved QBF*	$\bar{a}$:	667,906	596,840	-10.6
	$\bar{b}$:	6,104	4,549	-25.4		$\bar{b}$:	6,074	5,307	-12.6
	$\bar{t}$:	4.60	2.34	-49.0		$\bar{t}$:	4.75	2.81	-40.8
	$\tilde{a}$:	828	833	+0.6		$\tilde{a}$:	827	828	+0.1
	$\tilde{b}$:	1	1	+0.0		$\tilde{b}$:	1	1	+0.0
	$\tilde{t}$:	1.01	0.36	-63.8		$\tilde{t}$:	1.31	0.37	-71.7

to the planning instance. The performance of DepQBF-pre and Nenofex shown in Figure 4 illustrates the benefits of preprocessing in the planning workflow. Among other techniques, Bloqqer applies expansion, the core technique used in Nenofex, in a way that restricts the blow up of the formula size [4,7].

Given the results shown in Figure 4, preprocessing might considerably improve the performance of incremental QBF solving in our workflow. To this end, it is necessary to combine QBF preprocessing and solving in an *incremental* way.

5.2 Comparison to Heuristic Approaches

Although our focus is on a comparison of non-incremental and incremental QBF solving, we report on additional experiments with the heuristic planning tools ConformantFF [15] and T0 [30]. In contrast to our implemented QBF-based approach to conformant planning, heuristic tools do not guarantee to find a plan with the optimal (i.e., shortest) length. In practical settings, plans with optimal length are desirable. Moreover, the QBF-based approach allows to verify the non-existence of a plan with respect to an upper bound on the plan length. Due to these differences, a comparison based on the run times and numbers of solved instances only is not appropriate.

Related to Table 1, ConformantFF solved 169 planning instances, where it found a plan for 144 instances and concluded that no plan exists (with a length shorter than our considered upper bound of 200) for 25 instances. Considering the 124 instances where both incDepQBF and ConformantFF found a plan, for 42 instances the optimal plan found by incDepQBF was strictly shorter than the plan found by ConformantFF. On the

124 instances, the average (median) length of the plan found by incDepQBF was 2.06 (1), compared to an average (median) length of 3.45 (1) by ConformantFF.

Due to technical problems, we were not able to run the experiments with T0[4] on the same system as the experiments with (inc)DepQBF and ConformantFF. Hence the results by T0 reported in the following are actually incomparable to Table 1. However, we include them here to allow for a basic comparison of the plan lengths.

Using the same time and memory limits as for incDepQBF and ConformantFF, T0 solved 206 planning instances, where it found a plan for 203 instances and concluded that no plan exists (with a length shorter than the upper bound of 200) for three instances. Given the 156 instances where both incDepQBF and T0 found a plan, for 56 instances the optimal plan found by incDepQBF was strictly shorter than the plan found by T0. On the 156 instances, the average (median) length of the plan found by incDepQBF was 2.25 (1), compared to an average (median) length of 3.08 (2) by T0.

From the 13 instances solved by incDepQBF for which no plan exists (Table 1), none was solved using T0 and 12 were solved using ConformantFF.

Our experiments confirm that the QBF-based approach to conformant planning finds optimal plans in contrast to the plans found by the heuristic approaches implemented in ConformantFF and T0. Moreover, (inc)DepQBF and other search-based QBF solvers rely on Q-resolution [17] as the underlying proof system. Given a Q-resolution proof Π of the unsatisfiability of a QBF ψ, it is possible to extract from Π a countermodel [2] or strategy [13] of ψ in terms of a set of Herbrand functions. Intuitively, an Herbrand function $f_y(x_{y_1}, \ldots, x_{y_n})$ represents the values that a universal variable f_y must take to falsify ψ with respect to the values of all existential variables $x_{y_1}, \ldots, x_{y_n}$ with $x_{y_i} < y$ in the prefix ordering. Given a conformant planning problem P, Q-resolution proofs and Herbrand function countermodels allow to independently explain and verify [29] the non-existence of a plan (of a particular length) for P by verifying the unsatisfiability of the QBF encoding of P. This is an appealing property of the QBF-based approach. In practical applications, it may be interesting to have an explanation of the non-existence of a plan in addition to the mere answer that no plan exists.

The exact QBF-based approach for conformant planning can be combined with heuristic approaches in a portfolio-style system, for example. Thereby, the two approaches are applied in parallel and independently from each other. This way, modern multi-core hardware can naturally be exploited.

6   Conclusion

We presented a case study of incremental QBF solving based on a workflow to incrementally encode planning problems into sequences of QBFs. Thereby, we focused on a general-purpose QBF solver. The incremental approach avoids some redundancy. First, parts of the QBF encodings of shorter plan lengths can be reused in the encodings of longer plan lengths. Second, the incremental QBF solver benefits from information that was learned from previously solved QBFs. Compared to heuristic approaches, the QBF-based approach has the advantage that it always finds the shortest plan and it allows to verify the non-existence of a plan by Q-resolution proofs.

---

[4] Experiments with T0 were run on AMD Opteron 6176 SE, 2.3 GHz, 64-bit Linux.

Using variants of the solver DepQBF, incremental QBF solving outperforms non-incremental QBF solving in the planning workflow in terms of solved instances and statistics like the number of backtracks, assignments, and run time. The results of our experimental study independently motivate the use of incremental QBF solving in applications other than planning. We implemented the Java interface DepQBF4J to integrate the solver DepQBF in our planning tool. This interface is extensible and can be combined with arbitrary Java applications.

The experiments revealed that keeping learned information in incremental QBF solving might be harmful if the heuristics of the solver are negatively influenced. Our observations merit a closer look on these heuristics when used in incremental solving. In general, the combination of preprocessing and incremental solving [14,16,19,26,27,28,32] could improve the performance of QBF-based workflows.

References

1. Audemard, G., Lagniez, J.M., Simon, L.: Improving Glucose for Incremental SAT Solving with Assumptions: Application to MUS Extraction. In: Järvisalo, M., Van Gelder, A. (eds.) SAT 2013. LNCS, vol. 7962, pp. 309–317. Springer, Heidelberg (2013)
2. Balabanov, V., Jiang, J.H.R.: Unified QBF certification and its applications. Formal Methods in System Design 41(1), 45–65 (2012)
3. Baral, C., Kreinovich, V., Trejo, R.: Computational Complexity of Planning and Approximate Planning in the Presence of Incompleteness. Artificial Intelligence 122(1-2), 241–267 (2000)
4. Biere, A.: Resolve and Expand. In: H. Hoos, H., Mitchell, D.G. (eds.) SAT 2004. LNCS, vol. 3542, pp. 59–70. Springer, Heidelberg (2005)
5. Biere, A., Lonsing, F., Seidl, M.: Blocked Clause Elimination for QBF. In: Bjørner, N., Sofronie-Stokkermans, V. (eds.) CADE 2011. LNCS, vol. 6803, pp. 101–115. Springer, Heidelberg (2011)
6. Blum, A., Furst, M.L.: Fast Planning Through Planning Graph Analysis. Artificial Intelligence 90(1-2), 281–300 (1997)
7. Bubeck, U., Kleine Büning, H.: Bounded Universal Expansion for Preprocessing QBF. In: Marques-Silva, J., Sakallah, K.A. (eds.) SAT 2007. LNCS, vol. 4501, pp. 244–257. Springer, Heidelberg (2007)
8. Cadoli, M., Schaerf, M., Giovanardi, A., Giovanardi, M.: An Algorithm to Evaluate Quantified Boolean Formulae and Its Experimental Evaluation. Journal of Automated Reasoning 28(2), 101–142 (2002)
9. Davis, M., Logemann, G., Loveland, D.: A Machine Program for Theorem-proving. Communications of the ACM 5(7), 394–397 (1962)
10. Eén, N., Sörensson, N.: Temporal Induction by Incremental SAT Solving. Electronic Notes in Theoretical Computer Science 89(4), 543–560 (2003)
11. Giunchiglia, E., Narizzano, M., Tacchella, A.: Clause/Term Resolution and Learning in the Evaluation of Quantified Boolean Formulas. Journal of Artificial Intelligence Research 26, 371–416 (2006)
12. Giunchiglia, E., Marin, P., Narizzano, M.: sQueezeBF: An Effective Preprocessor for QBFs Based on Equivalence Reasoning. In: Strichman, O., Szeider, S. (eds.) SAT 2010. LNCS, vol. 6175, pp. 85–98. Springer, Heidelberg (2010)
13. Goultiaeva, A., Van Gelder, A., Bacchus, F.: A Uniform Approach for Generating Proofs and Strategies for Both True and False QBF Formulas. In: Walsh, T. (ed.) IJCAI, pp. 546–553. AAAI Press (2011)

14. Heule, M., Seidl, M., Biere, A.: A Unified Proof System for QBF Preprocessing. In: Demri, S., Kapur, D., Weidenbach, C. (eds.) IJCAR 2014. LNCS, vol. 8562, pp. 91–106. Springer, Heidelberg (2014)
15. Hoffmann, J., Brafman, R.I.: Conformant planning via heuristic forward search: A new approach. Artificial Intelligence 170(6-7), 507–541 (2006)
16. Janota, M., Grigore, R., Marques-Silva, J.: On QBF Proofs and Preprocessing. In: McMillan, K., Middeldorp, A., Voronkov, A. (eds.) LPAR-19 2013. LNCS, vol. 8312, pp. 473–489. Springer, Heidelberg (2013)
17. Kleine Büning, H., Karpinski, M., Flögel, A.: Resolution for Quantified Boolean Formulas. Information and Computation 117(1), 12–18 (1995)
18. Kronegger, M., Pfandler, A., Pichler, R.: Conformant planning as a benchmark for QBF-solvers. In: Report Int. Workshop on Quantified Boolean Formulas (QBF 2013). pp. 1–5 (2013), http://fmv.jku.at/qbf2013/reportQBFWS13.pdf
19. Kupferschmid, S., Lewis, M.D.T., Schubert, T., Becker, B.: Incremental Preprocessing Methods for Use in BMC. Formal Methods in System Design 39(2), 185–204 (2011)
20. Lagniez, J.M., Biere, A.: Factoring Out Assumptions to Speed Up MUS Extraction. In: Järvisalo, M., Van Gelder, A. (eds.) SAT 2013. LNCS, vol. 7962, pp. 276–292. Springer, Heidelberg (2013)
21. Letz, R.: Lemma and Model Caching in Decision Procedures for Quantified Boolean Formulas. In: Egly, U., Fermüller, C. (eds.) TABLEAUX 2002. LNCS (LNAI), vol. 2381, pp. 160–175. Springer, Heidelberg (2002)
22. Lonsing, F., Biere, A.: Nenofex: Expanding NNF for QBF Solving. In: Kleine Büning, H., Zhao, X. (eds.) SAT 2008. LNCS, vol. 4996, pp. 196–210. Springer, Heidelberg (2008)
23. Lonsing, F., Egly, U.: Incremental QBF Solving. In: O'Sullivan, B. (ed.) CP 2014. LNCS, vol. 8656, pp. 514–530. Springer, Heidelberg (2014)
24. Lonsing, F., Egly, U.: Incremental QBF Solving by DepQBF. In: Hong, H., Yap, C. (eds.) ICMS 2014. LNCS, vol. 8592, pp. 307–314. Springer, Heidelberg (2014)
25. Lonsing, F., Egly, U., Van Gelder, A.: Efficient Clause Learning for Quantified Boolean Formulas via QBF Pseudo Unit Propagation. In: Järvisalo, M., Van Gelder, A. (eds.) SAT 2013. LNCS, vol. 7962, pp. 100–115. Springer, Heidelberg (2013)
26. Marin, P., Miller, C., Becker, B.: Incremental QBF Preprocessing for Partial Design Verification - (Poster Presentation). In: Cimatti, A., Sebastiani, R. (eds.) SAT 2012. LNCS, vol. 7317, pp. 473–474. Springer, Heidelberg (2012)
27. Marin, P., Miller, C., Lewis, M.D.T., Becker, B.: Verification of Partial Designs using Incremental QBF Solving. In: Rosenstiel, W., Thiele, L. (eds.) DATE, pp. 623–628. IEEE (2012)
28. Nadel, A., Ryvchin, V., Strichman, O.: Ultimately Incremental SAT. In: Sinz, C., Egly, U. (eds.) SAT 2014. LNCS, vol. 8561, pp. 206–218. Springer, Heidelberg (2014)
29. Niemetz, A., Preiner, M., Lonsing, F., Seidl, M., Biere, A.: Resolution-Based Certificate Extraction for QBF - (Tool Presentation). In: Cimatti, A., Sebastiani, R. (eds.) SAT 2012. LNCS, vol. 7317, pp. 430–435. Springer, Heidelberg (2012)
30. Palacios, H., Geffner, H.: Compiling Uncertainty Away in Conformant Planning Problems with Bounded Width. Journal of Artificial Intelligence Research 35, 623–675 (2009)
31. Rintanen, J.: Asymptotically Optimal Encodings of Conformant Planning in QBF. In: Holte, R.C., Howe, A.E. (eds.) AAAI, pp. 1045–1050. AAAI Press (2007)
32. Seidl, M., Könighofer, R.: Partial witnesses from preprocessed quantified Boolean formulas. In: DATE, pp. 1–6. IEEE (2014)
33. Smith, D.E., Weld, D.S.: Conformant Graphplan. In: Mostow, J., Rich, C. (eds.) AAAI/IAAI, pp. 889–896. AAAI Press / The MIT Press (1998)
34. Zhang, L., Malik, S.: Towards a Symmetric Treatment of Satisfaction and Conflicts in Quantified Boolean Formula Evaluation. In: Van Hentenryck, P. (ed.) CP 2002. LNCS, vol. 2470, pp. 200–215. Springer, Heidelberg (2002)

Dynamic Symmetry Breaking in Itemset Mining

Belaïd Benhamou[1,2,*]

[1] Aix-Marseille Université
Laboratoire des Sciences de l'information et des Systèmes (LSIS)
Domaine universitaire de Saint Jérôme, Avenue Escadrille Normandie Niemen 13397
MARSEILLE Cedex 20
[2] Centre de Recherche en Informatique de Lens (CRIL)
Université d'Artois, Rue Jean Souvenir, SP 18 F 62307 Lens Cedex
belied.benhamou@univ-amu.fr

Abstract. The concept of symmetry has been extensively studied in the field of constraint programming and in the propositional satisfiability. We are interested here, by the detection and elimination of local and global symmetries in the item-set mining problem. Recent works have provided effective encodings as Boolean constraints for these data mining tasks and some idea on symmetry elimination in this area begin to appear, but still few and the techniques presented are often on *global symmetry* that is detected and eliminated statically in a preprocessing phase. In this work we study the notion of *local symmetry* and compare it to *global symmetry* for the itemset mining problem. We show how local symmetries of the boolean encoding can be detected dynamically and give some properties that allow to eliminate theses symmetries in SAT-based itemset mining solvers in order to enhance their efficiency.

Keywords: symmetry, Item-set mining, data mining, satisfiability, constraint programming.

1 Introduction

The work we propose here is to investigate the notion of *local symmetry*[1] elimination in the Frequent Itemset Mining (FIM) [1] and compare it to *global symmetry*[2]. The itemset mining problem has several applications in real-life problems and remains central in the Data mining research field. Since its introduction in 1993 [1], several highly scalable algorithms are introduction ([2], [20], [39],[36] [37], [13],[16], [31]) to enumerate the sets of frequent items.

Recently DeRaedt et Al. ([33], [18]) introduced the alternative of using constraint programming in data mining. They showed that a such alternative can be efficiently applied for a wide range of pattern mining problems. Most of the pattern mining constraint had been expressed in a declarative constraint programming language. This include frequency constraint, closeness, maximality, and anti-monotonic then use a constraint programming system like Gecode as black box to solve the problem. A strength point here

---

* Actually, I am at CRIL for one year CNRS delegation position.
[1] The symmetry of the sub-problems corresponding the different nodes of the search tree.
[2] The symmetry of the initial problem corresponding to the root of the search tree.

G.A. Aranda-Corral et al. (Eds.): AISC 2014, LNAI 8884, pp. 132–146, 2014.
© Springer International Publishing Switzerland 2014

is that different constraints can be combined and solved without the need to modify the solver, unlike in the existing specific data mining algorithms. Since the introduction of this declarative approach, there is a growing interest in finding generic methods to solve data mining tasks. For instance, several works expressed data mining problems as boolean satisfiability problem ([24], [21], [29], [26], [34], [23]) and used efficient modern SAT solvers as black box to solve them. More recently, a constraint declarative framework for solving Data mining tasks called MininZinc [17], had been introduced.

On the other hand, symmetry is a fundamental property that can be used to study various complex objects, to finely analyze their structures or to reduce the computational complexity when dealing with combinatorial problems. Krishnamurthy introduced in [28] the principle of symmetry to improve resolution in propositional logic. Symmetries for Boolean constraints are studied in depth in [8, 9]. The authors showed how to detect them and proved that their exploitation is a real improvement for several automated deduction algorithms. Since that, many research works on symmetry appeared. For instance, the static approach used by James Crawford et al. in [14] for propositional logic theories consists in adding constraints expressing global symmetry of the problem. This technique has been improved in [6] and extended to 0-1 Integer Logic Programming in [4]. But the notion of symmetry in the field of data mining is not well studied yet. Only few works on *global symmetry* elimination are introduced for some specific data mining algorithms that are targeted to solve some Data mining tasks ([19], [30],[15], [38], [32], [25],[22]).

As far as we know, there is no *local symmetry* breaking method in the framework of data mining. In this work, we investigate dynamic local symmetry detection and elimination and compare to *global symmetry* exploitation in SAT-based item set mining solvers. *Local symmetry* is the symmetry that we can discover at each node of the search tree during search. *Global symmetry* is the particular local symmetry corresponding to the root of the search tree (the symmetry of the initial problem). Almost all of the known works on symmetry are on global symmetry. Only few works on local symmetry [8, 9] are known in the literature. Local symmetry breaking remains a big challenge.

Eliminating symmetry leads to enumerate only the non symmetrical structures, then could provide a more pertinent and compact output. That is only non-symmetrical patterns are generated, each symmetrical pattern class is represented by one element.

The rest of the paper is structured as follows: in Section 2, we give some necessary background on the satisfiability problem, permutations and the necessary notion on itemset mining problem. We study the notion of symmetry in itemset mining in Section 3. In Section 4 we show how symmetry can be detected by means of graph automorphism. We show how local and global symmetry can be eliminated in Section 5. Section 6 shows how symmetry elimination is exploited by a SAT-based item set mining solvers and we gives some experiments on different transaction data-sets in section 7. We conclude the work in Section 8.

2   Background

We summarize in this section some background on the satisfiability problem, permutations, and the necessary notions on the itemset mining problem.

2.1   The Propositional Satisfiability Problem (SAT)

We shall assume that the reader is familiar with the propositional calculus. We give here, a short description. Let V be the set of propositional variables called only variables. Variables will be distinguished from literals, which are variables with an assigned parity 1 or 0 that means $True$ or $False$, respectively. This distinction will be ignored whenever it is convenient, but not confusing. For a propositional variable p, there are two literals: p the positive literal and $\neg p$ the negative one.

A clause is a disjunction of literals $\{p_1, p_2, \ldots, p_n\}$ such that no literal appears more than once, nor a literal and its negation at the same time. This clause is denoted by $p_1 \vee p_2 \vee \ldots \vee p_n$. A set $\mathcal{F}$ of clauses is a conjunction of clauses. In other words, we say that $\mathcal{F}$ is in the conjunctive normal form (CNF).

A truth assignment of a system of clauses $\mathcal{F}$ is a mapping I defined from the set of variables of $\mathcal{F}$ into the set $\{True, False\}$. If $I[p]$ is the value for the positive literal p then $I[\neg p] = 1 - I[p]$. The value of a clause $p_1 \vee p_2 \vee \ldots \vee p_n$ in I is $True$, if the value $True$ is assigned to at least one of its literals in I, $False$ otherwise. By convention, we define the value of the empty clause ($n = 0$) to be $False$. The value $I[\mathcal{F}]$ of the system of clauses is $True$ if the value of each clause of $\mathcal{F}$ is $True$, $False$, otherwise. We say that a system of clauses $\mathcal{F}$ is satisfiable if there exists some truth assignments I that assign the value $True$ to $\mathcal{F}$, it is unsatisfiable otherwise. In the first case I is called a model of $\mathcal{F}$. Let us remark that a system which contains the empty clause is unsatisfiable.

It is well-known [35] that for every propositional formula $\mathcal{F}$ there exists a formula $\mathcal{F}'$ in conjunctive normal form(CNF) such that $\mathcal{F}'$ is satisfiable iff $\mathcal{F}$ is satisfiable. In the following we will assume that the formulas are given in a conjunctive normal form.

2.2   Permutations

Let $\Omega = \{1, 2, \ldots, N\}$ for some integer N, where each integer might represent a propositional variable or an atom. A permutation of Ω is a bijective mapping σ from Ω to Ω that is usually represented as a product of cycles of permutations. We denote by $Perm(\Omega)$ the set of all permutations of Ω and $\circ$ the composition of the permutation of $Perm(\Omega)$. The pair $(Perm(\Omega), \circ)$ forms the permutation group of Ω. That is, $\circ$ is closed and associative, the inverse of a permutation is a permutation and the identity permutation is a neutral element. A pair $(T, \circ)$ forms a sub-group of $(S, \circ)$ iff T is a subset of S and forms a group under the operation $\circ$.

The orbit $\omega^{Perm(\Omega)}$ of an element ω of Ω on which the group $Perm(\Omega)$ acts is $\omega^{Perm(\Omega)} = \{\omega^\sigma : \omega^\sigma = \sigma(\omega), \sigma \in Perm(\Omega)\}$.

A generating set of the group $Perm(\Omega)$ is a subset Gen of $Perm(\Omega)$ such that each element of $Perm(\Omega)$ can be written as a composition of elements of Gen. We write $Perm(\Omega) = < Gen >$. An element of Gen is called a generator. The orbit of $\omega \in \Omega$ can be computed by using only the set of generators Gen.

2.3   The Frequent, Closed, Maximal Itemset Problem

Let $\mathcal{I} = \{0, \ldots, m - 1\}$ be a set of m items and $\mathcal{T} = \{0, \ldots, n - 1\}$ a set of n transactions (transaction identifier). A subset $I \subseteq \mathcal{I}$ is called an itemset and a transaction $t \in \mathcal{T}$ over $\mathcal{I}$ is in fact, a pair (t_{id}, I) where t_{id} is the transaction identifier and I the corresponding itemset. Usually, when there is no confusion, a transaction is just expressed by its identifier. A transaction database $\mathcal{D}$ over $\mathcal{I}$ is a finite set of transactions such that no different transactions have the same identifier. A transaction database can be seen as a binary matrix $n \times m$, where $n =| \mathcal{T} |$ and $m =| \mathcal{I} |$, with $\mathcal{D}_{t,i} \in \{0,1\}$ forall $t \in \mathcal{T}$ and $i \in \mathcal{I}$. More precisely, a transaction database is expressed by the set $D = \{(t, I) \mid t \in \mathcal{T}, I \subseteq \mathcal{I}, \forall i \in I : \mathcal{D}_{t,i} = 1\}$. The *coverage* $C_{\mathcal{D}}(I)$ of an itemset I in $\mathcal{D}$ is the set of all transactions in which I occurs. That is, $C_{\mathcal{D}}(I) = \{t \in \mathcal{T} \mid \forall i \in I, \mathcal{D}_{t,i} = 1\}$. The *support* $S_{\mathcal{D}}(I)$ of I in $\mathcal{D}$ is the number $| C_{\mathcal{D}}(I) |$ of transactions supporting I. Moreover, the *frequency* $F_{\mathcal{D}}(I)$ of I in $\mathcal{D}$ is defined by $\frac{|C_{\mathcal{D}}(I)|}{|\mathcal{D}|}$.

Example 1. Consider the transaction database $\mathcal{D}$ made over the set of fruit items $\mathcal{I} = \{Kiwi, Oranges, Apple, Cherries, plums\}$. For example, we can see in Table 1 that the itemset $I = \{kiwi, Apples\}$ has $C_{\mathcal{D}}(I) = \{1,3\}$, $S_{\mathcal{D}}(I) =| C_{\mathcal{D}}(I) |= 2$, and $F_{\mathcal{D}}(I) = 0, 5$.

Table 1. An instance of a transaction database

t_{id}	itemset
1	Cherries, Apples, Kiwi
2	Cherries, Apples, Oranges
3	Plums, Apples, Kiwi
4	Plums, Apples, Oranges

Given a transaction database $\mathcal{D}$ over $\mathcal{L}$, and θ a minimal support threshold, an itemset I is said to be frequent if $S_{\mathcal{D}}(I) \geq \theta$. I is a closed frequent itemset if in addition to the frequency constraint it satisfies the following constraint: for all itemset J such that $I \subset J, S_{\mathcal{D}}(I) > S_{\mathcal{D}}(J)$. I is said to be a maximal frequent itemset if in addition to the frequency constraint it satisfies the following constraint: for all itemset J such that $I \subset J, S_{\mathcal{D}}(J) < \theta$. Both closed and maximal itemsets are two known condensed representation for frequent itemsets. The data mining tasks we are dealing with in this work are defined as follows:

Definition 1. *1. The frequent itemset mining task consists in computing the following set $\mathcal{FIM}_{\mathcal{D}}(\theta) = \{I \subseteq \mathcal{I} | S_{\mathcal{D}}(I) \geq \theta\}$.*

2. The closed frequent itemset mining task consists in computing the following set $\mathcal{CLO}_{\mathcal{D}}(\theta) = \{I \in \mathcal{FIM}_{\mathcal{D}}(\theta) | \forall J \subseteq \mathcal{I}, I \subset J, S_{\mathcal{D}}(I) > S_{\mathcal{D}}(J)\}$.

In the next section, we will use the previous definition to express both the frequent and the closed frequent itemset mining tasks as declarative constraints that could be solved by appropriate constraint solvers.

3 Symmetry in Itemset Mining Represented as a Satisfiability Problem

The frequent itemset mining tasks and some of its variants tasks (closed, maximal, ..etc) had been encoded for the first time in [33, 18] as constraint programming tasks where a constraint solver could be used as a black box to solve them. Since that, other works ([24], [21], [29], [26], [34], [23]) expressed the data mining tasks as a satisfiability problem where the mining tasks are represented by propositional formulas that are translated into their conjunctive normal forms (CNF) which will be given as inputs to a SAT solver. In this work we use this last approach to encode the data mining tasks as satisfiability problems in which we detect and eliminate symmetry.

Before defining symmetry, we shall first give the CNF encoding of the data mining tasks. The idea behind the CNF encoding of a data mining task defined on a database transaction database $\mathcal{D}$ is to express each of its interpretations as a pair (I, T) where I represents an itemset and T its covering transaction subset in $\mathcal{D}$. To do that, a boolean variable I_i is associated with each item $i \in \mathcal{I}$ and a variable T_t is associated with each transaction $t \in \mathcal{T}$. The itemset I is then defined by all the variables I_i that are true. That is $I_i = 1$, if $i \in I$, and $I_i = 0$ if $i \notin I$. The set of transaction T covered by I is then defined by the set of variable T_t that are true. That is, $T_t = 1$ if $t \in C_\mathcal{D}(I)$ and $T_t = 0$ if $t \notin C_\mathcal{D}(I)$.

For instance, the $\mathcal{FIM}_\mathcal{D}(\theta)$ task can be seen as the search of the set of models $M = \{(I, T) \mid I \subseteq \mathcal{I}, T \subseteq \mathcal{T}, T = C_\mathcal{D}(I), \mid T \mid \geq \theta\}$. We have to encode both the covering constraint $T = C_\mathcal{D}(I)$ and the frequency constraint $\mid T \mid \geq \theta$. These constraints expressed by the two following boolean and pseudo boolean constraints :

$$\bigwedge_{t \in \mathcal{T}} (\neg T_t \leftarrow \bigvee_{i \in \mathcal{I}, \mathcal{D}_{t,i}=0} I_i)$$

$$\bigwedge_{i \in \mathcal{I}} (I_i \rightarrow \sum_{t \in \mathcal{T} \mid \mathcal{D}_{t,i}=1} T_t \geq \theta)$$

The frequent closed itemset task is specified by adding to the two previous constraints the two following constraint :

$$\bigwedge_{t \in \mathcal{T}} (\neg T_t \rightarrow \bigvee_{i \in \mathcal{I}, \mathcal{D}_{t,i}=0} I_i)$$

$$\bigwedge_{i \in \mathcal{I}} (I_i \leftrightarrow \bigwedge_{t \in \mathcal{T} \mid \mathcal{D}_{t,i}} T_t))$$

An important property of these logical encodings established to represent different data mining tasks is that the models of the resulting logical formulas express the solutions of the original data mining tasks considered. This approach is totally declarative, the logical formulas representing the data mining tasks are translated to their equivalent CNF formulas by using known transformation techniques [35] and then given as inputs to a SAT solver which is used as a black box to compute theirs models. For example, if the considered problem is the search of frequent itemsets in a transaction database $\mathcal{D}$, then the models of the logical formula representing this task in $\mathcal{D}$ express exactly the different frequent itemsets of $\mathcal{D}$ and their covers. That is, if $CNF(k, \mathcal{D})$ denotes the CNF logic encoding a the data mining task k in the transaction database $\mathcal{D}$ and $P_\mathcal{D}^k$ a predicate representing the task k in $\mathcal{D}$, then an itemset $I \subseteq \mathcal{I}$ having $T \subseteq \mathcal{T}$ as a cover verifies $P_\mathcal{D}^k$ ($P_\mathcal{D}^k(I, T) = true$) if I is an itemset which is an answer to the data mining task k and T is its cover. Formally, we get the following proposition:

Proposition 1. *Let $J = (I, T)$ be an interpretation of $CNF(k, \mathcal{D})$, $I' = \{i \in \mathcal{I} : I_i = true\}$, and $T' = \{t \in \mathcal{T} : T_t = true\}$, then J is a model of $CNF(k, \mathcal{D})$ iff $P_{\mathcal{D}}^k(I', T') = true$.*

Proof. The proof is similar to that one given in [33, 18]. It expresses the fact that the boolean encoding $CNF(k, \mathcal{D})$ is sound.

On other hand symmetry is well studied in constraint programming and in the satisfiability problem. Since Krishnamurthy's [27] symmetry definition and the one given in [11, 12] in propositional logic, several other definitions are given in the CP community. We will define in the following both semantic and syntactic symmetries for the boolean encoding of the itemset mining problem and show their relationship.

Definition 2. (Semantic Symmetry of $CNF(k, \mathcal{D})$) *Let $CNF(k, \mathcal{D})$ be the CNF encoding of the mining task k in $\mathcal{D}$ and $L_{CNF(k,\mathcal{D})}$ its set of literals. A semantic symmetry of $CNF(k, \mathcal{D})$ is a permutation σ defined on $L_{CNF(k,\mathcal{D})}$ such that $CNF(k, \mathcal{D})$ and $\sigma(CNF(k, \mathcal{D}))$ have the same models (i.e. $\mathcal{D}$ and $\sigma(\mathcal{D})$ have the same frequent patterns).*

In other words a semantic symmetry of $CNF(k, \mathcal{D})$ is a literal permutation that conserves the set of frequent/closed or maximal item sets of $\mathcal{D}$. Semantic symmetry is a general symmetry definition, but its computation is trivially time consuming. We give in the following the definition of syntactic symmetry which we will show that it is a sufficient condition to semantic symmetry that could be computed efficiently.

Definition 3. (Syntactic Symmetry of $CNF(k, \mathcal{D})$) *Let $CNF(k, \mathcal{D})$ be the boolean encoding of the data mining task k defined on $\mathcal{D}$ and $L_{CNF(k,\mathcal{D})}$ its set of literals. A syntactic symmetry of $CNF(k, \mathcal{D})$ is a permutation σ defined on $L_{CNF(k,\mathcal{D})}$ such that the following conditions hold:*

1. $\forall \ell \in L_{CNF(k,\mathcal{D})}, \sigma(\neg \ell) = \neg \sigma(\ell)$,
2. $\sigma(CNF(k, \mathcal{D})) = CNF(k, \mathcal{D})$

In other words, a syntactical symmetry of $CNF(k, \mathcal{D})$ is a literal permutation that leaves $CNF(k, \mathcal{D})$ invariant. If we denote by $Perm(L_{CNF(k,\mathcal{D})})$ the group of permutations of $L_{CNF(k,\mathcal{D})}$ and by $Sym(L_{CNF(k,\mathcal{D})}) \subseteq Perm(L_{CNF(k,\mathcal{D})})$ the subset of permutations of $L_{CNF(k,\mathcal{D})}$ that are the syntactic symmetries of $CNF(k, \mathcal{D})$, then $Sym(L_{CNF(k,\mathcal{D})}$ is trivially a sub-group of $Perm(L_{CNF(k,\mathcal{D})})$.

Theorem 1. *Each syntactical symmetry of $CNF(k, \mathcal{D})$ is a semantic symmetry of $CNF(k, \mathcal{D})$.*

Proof. It is trivial to see that a syntactic symmetry of $CNF(k, \mathcal{D})$ is always a semantic symmetry of $CNF(k, \mathcal{D})$. Indeed, if σ is a syntactic symmetry of $CNF(k, \mathcal{D})$, then $\sigma(CNF(k, \mathcal{D})) = CNF(k, \mathcal{D})$, thus it results that $CNF(k, \mathcal{D})$ and $\sigma(CNF(k, \mathcal{D}))$ have the same models (they express the same item sets satisfying the predicate $P_{\mathcal{D}}^k$)).

Example 2. Consider the transaction database $\mathcal{D}$ of Table 1 and $k = \mathcal{FIM}_{\mathcal{D}}(\theta)$ for $\theta = 2$. If the set of items $\mathcal{I} = \{Kiwi, Oranges, Apple, Cherries, Plums\}$ are encoded by the scalars $\{1, 2, 3, 4, 5\}$, then the corresponding boolean encoding

$CNF(\mathcal{FI}\text{-}\mathcal{M}_\mathcal{D}(\theta), \mathcal{D})$ for the frequent item set mining in $\mathcal{D}$ is formed by the set $var = \{I_1, I_2, I_3, I_4, I_5, T_1, T_2, T_3, T_4\}$ of boolean variables and the set $cl = \{\neg T_1 \vee \neg I_2, \neg T_1 \vee \neg I_5, \neg T_2 \vee \neg I_1, \neg T_2 \vee \neg I_5, \neg T_3 \vee \neg I_2, \neg T_3 \vee \neg I_4, \neg T_4 \vee \neg I_1, \neg T_4 \vee \neg I_4\}$ of clauses and the pseudo boolean constraints $pb = \{I_1 \rightarrow T_1 + T_3 \geq 2, I_2 \rightarrow T_2 + T_4 \geq 2, I_3 \rightarrow T_1 + T_2 + T_3 + T_4 \geq 2, I_4 \rightarrow T_1 + T_2 \geq 2, I_5 \rightarrow T_3 + T_4 \geq 2\}$. The permutation $\sigma = (I_1, I_2)(I_4, I_5)(T_1, T_4)(T_2, T_3)$ defined on the set of variables var is a syntactic symmetry of $CNF(\mathcal{FI}\mathcal{M}_\mathcal{D}(\theta), \mathcal{D})$.

In the sequel we give some symmetry properties of the boolean encoding $CNF(k, \mathcal{D})$, which express some semantics on the database $\mathcal{D}$.

Definition 4. *Two literals I_i and I_j of $CNF(k, \mathcal{D})$ are symmetrical if there exists a symmetry σ of $CNF(k, \mathcal{D})$ such that $\sigma(I_i) = I_j$.*

Remark 1. The symmetry between the item literals I_i and I_j expresses the symmetry between the items i and j of $\mathcal{D}$. The previous definition could be applied for the transaction literals T_t too.

Definition 5. *The orbit of a literal $I_i \in CNF(k, \mathcal{D})$ on which the group of symmetries $Sym(L_{CNF(k,\mathcal{D})})$ acts is $I_i^{Sym(L_{CNF(k,\mathcal{D})})} = \{\sigma(I_i) : \sigma \in Sym(L_{CNF(k,\mathcal{D})})\}$*

Remark 2. All the literals in the orbit of a literal I_i are symmetrical two by two.

Example 3. In Example 2, the orbit of the item I_1 is $I_1^{Sym(L_{CNF(k,\mathcal{D})})} = \{I_1, I_2\}$,

If I is a model of $CNF(k, \mathcal{D})$ and σ a syntactic symmetry, we can get another model of $CNF(k, \mathcal{D})$ by applying σ on the literals which appear in I. Formally we get the following property. These two symmetrical models of $CNF(k, \mathcal{D})$ express two symmetrical item sets of $\mathcal{D}$.

Proposition 2. *I is a model of $CNF(k, \mathcal{D})$ iff $\sigma(I)$ is a model of $CNF(k, \mathcal{D})$.*

Proof. Suppose that I is a model of $CNF(k, \mathcal{D})$, then $\sigma(I)$ is a model of $\sigma(CNF(k, \mathcal{D}))$. We can then deduce that $\sigma(I)$ is a model of $CNF(k, \mathcal{D})$ since $CNF(k, \mathcal{D})$ is invariant under σ. The converse can be shown by considering the converse permutation of σ.

In Example 2, if we consider $\theta = 2$ and the symmetry σ, there will be symmetrical models in $CNF(k, \mathcal{D})$ (symmetrical frequent item sets in $\mathcal{D}$). For instance, $J = (I, T) = \{I_1, I_3, T_1, T_3\}$ is a model of $CNF(k, \mathcal{D})$ that corresponds to the frequent item set $\{Kiwi, Apples\}$ in $\mathcal{D}$. By the symmetry σ we can deduce that $\sigma(J) = \{I_2, I_3, T_2, T_4\}$ is also a model of $CNF(k, \mathcal{D})$ which corresponds to the frequent item sets $\{Oranges, Apples\}$. These are what we call symmetrical models of $CNF(k, \mathcal{D})$ or symmetrical frequent item sets of $\mathcal{D}$. A symmetry σ transforms each frequent itemset (a model of the CNF encoding) into a frequent itemset and each no-good (not a frequent itemset or a model of the CNF encoding) into a no-good.

Theorem 2. *Let I_i and I_j be two literals of $CNF(k, \mathcal{D})$ that are in the same orbit with respect to the symmetry group $Sym(L_{CNF(\mathcal{D})})$, then I_i is true in an a model of $CNF(k, \mathcal{D})$ iff I_j is true in a model of $CNF(k, \mathcal{D})$.*

Proof. If I_i is in the same orbit as I_j then it is symmetrical with I_j in $CNF(k, \mathcal{D})$. Thus, there exists a symmetry σ of $CNF(k, \mathcal{D})$ such that $\sigma(I_i) = I_j$. If I is a model of $CNF(k, \mathcal{D})$ then $\sigma(I)$ is also a model of $\sigma(CNF(k, \mathcal{D})) = CNF(k, \mathcal{D})$, besides if $I_i \in I$ then $I_j \in \sigma(I)$ which is also a model of $CNF(k, \mathcal{D})$. For the converse, consider $I_i = \sigma^{-1}(I_j)$, and make a similar proof.

Corollary 1. *Let I_i be a literal of $CNF(k, \mathcal{D})$, if I_i is not true in any model of $CNF(k, \mathcal{D})$, then each literal $I_j \in orbit^\ell = \ell^{Sym(L_{CNF(k,\mathcal{D})})}$ is not true in any model of $CNF(k, \mathcal{D})$.*

Proof. The proof is a direct consequence of Theorem 2.

Corollary 1 expresses an important property that we will use to break local symmetry at each node of the search tree of a SAT-based procedure for the itemset mining problem. That is, if a no-good is detected after assigning the value True to the current literal I_i of $CNF(k, \mathcal{D})$, then we compute the orbit of I_i and assign the value false to each literal in it, since by symmetry the value true will not lead to any model of $CNF(k, \mathcal{D})$.

4 Symmetry Detection

The most known technique to detect syntactic symmetries for CNF formulas in satisfiability is the one consisting in reducing the considered formula into a graph [14, 5, 4] whose automorphism group is identical to the symmetry group of the original formula. We adapt the same approach here to detect the syntactic symmetries of the boolean encoding $CNF(k, \mathcal{D})$ of transaction database. That is, we represent the boolean encoding $CNF(k, \mathcal{D})$ of the transaction database $\mathcal{D}$ by a graph $G_{CNF(k,\mathcal{D})}$ that we use to compute the symmetry group of $CNF(k, \mathcal{D})$ by means of its automorphism group. When this graph is built, we use a graph automorphism tool like Saucy [5] to compute its automorphism group which gives the symmetry group of $CNF(k, \mathcal{D})$. Following the technique used in [14, 5, 4], we summarize bellow the construction of the graph which represent the boolean encoding $CNF(k, \mathcal{D})$. Given the the encoding $CNF(k, \mathcal{D})$, the associated colored graph $G_{CNF(k,\mathcal{D})}(V, E)$ is defined as follows:

- Each positive item literal I_i of $CNF(k, \mathcal{D})$ is represented by a vertex $I_i \in V$ of the color 1 in $G_{CNF(k,\mathcal{D})}$. The negative literal $\neg I_i$ associated with I_i is represented by a vertex $\neg I_i$ of color 1 in $G_{CNF(k,\mathcal{D})}$. These two literal vertices are connected by an edge of E in the graph $G_{CNF(k,\mathcal{D})}$.
- Each positive transaction literal T_t of $CNF(k, \mathcal{D})$ is represented by a vertex $T_t \in V$ of the color 2 in $G_{CNF(k,\mathcal{D})}$. The negative literal $\neg T_t$ associated with T_t is represented by a vertex $\neg T_t$ of color 2 in $G_{CNF(k,\mathcal{D})}$. These two literal vertices are connected by an edge of E in the graph $G_{CNF(k,\mathcal{D})}$.
- Each positive auxiliary[3] literal ℓ_i of $CNF(k, \mathcal{D})$ is represented by a vertex $\ell_i \in V$ of the color 3 in $G_{CNF(k,\mathcal{D})}$. The negative literal $\neg \ell_i$ associated with ℓ_i is represented by a vertex $\neg \ell_i$ of color 3 in $G_{CNF(k,\mathcal{D})}$. These two literal vertices are connected by an edge of E in the graph $G_{CNF(k,\mathcal{D})}$.

---

[3] The literals used to compute the CNF form $CNF(k, \mathcal{D})$.

– Each clause c_i of $CNF(k, \mathcal{D})$ is represented by a vertex $c_i \in V$ (a clause vertex) of color 4 in $G_{CNF(k,\mathcal{D})}$. An edge connects this vertex c_i to each vertex representing one of its literals.

An important property of the graph $G_{CNF(k,\mathcal{D})}$ is that it preserves the syntactic group of symmetries of $CNF(k, \mathcal{D})$. That is, the syntactic symmetry group of $CNF(k, \mathcal{D})$ is identical to the automorphism group of its graph representation $G_{CNF(k,\mathcal{D})}$, thus we could use a graph automorphism system like Saucy on $G_{CNF(k,\mathcal{D})}$ to detect the syntactic symmetry group of $CNF(k, \mathcal{D})$. The graph automorphism system returns a set of generators Gen of the symmetry group from which we can deduce each symmetry of $CNF(k, \mathcal{D})$.

5   Symmetry Elimination

There are two ways to break symmetry. The first one is to deal with the global symmetry which is present in the formulation of the given problem. Global symmetry can be eliminated in a static way in a pre-processing phase of a SAT-based itemset solver by just adding the symmetry predicates as it is done in [14, 5, 6, 4]. The second way is the elimination of local symmetry that could appear in the sub-problems corresponding to the different nodes of the search tree of a SAT-based itemset solver. Global symmetry can be considered as the local symmetry corresponding to the root of the search tree.

Local symmetries have to be detected and eliminated dynamically at some decision node of the search tree. Dynamic symmetry detection in satisfiability had been studied in [8–10] where a local syntactic symmetry search method had been given. We use the same technique to break local symmetry in itemset mining.

Consider the logic encoding $CNF(k, \mathcal{D})$ of the transaction $\mathcal{D}$, and a partial assignment I of a SAT-based itemset solver applied to $CNF(k, \mathcal{D})$. Suppose that ℓ is the current literal under assignment. The assignment I simplifies $CNF(k, \mathcal{D})$ into a sub-formula $CNF(k, \mathcal{D})_I$ which defines a state in the search space corresponding to the current node n_I of the search tree. The main idea is to maintain dynamically the graph $G_{CNF(k,\mathcal{D})}$ of the sub-formula $CNF(k, \mathcal{D})_I$ corresponding to the current node n_I, then color the graph $G_{CNF(k,\mathcal{D})_I}$ as shown in the previous section and compute its automorphism group $Aut(CNF(k, \mathcal{D})_I)$. The sub-formula $CNF(k, \mathcal{D})_I$ can be viewed as the remaining sub-problem corresponding to the unsolved part. By applying an automorphism tool on this colored graph we can get the generator set Gen of the symmetry sub-group existing between literals from which we can compute the orbit of the current literal ℓ that we will use to make the symmetry cut.

After this, we use Corollary 1 to break dynamically the local symmetry and then prune search spaces of tree search itemset methods. Indeed, if the assignment of the current literal ℓ defined at a given node n_I of the search tree is shown to be a failure, then by symmetry, the assignment of each literal in the orbit of ℓ will result in a failure too. Therefore, the negated literal of each literal in the orbit of ℓ has to be assigned the value true in the partial assignment I. Thus, we prune in the search tree, the sub-space which corresponds to true assignment of the literals of the orbit of ℓ. That is what we call the local symmetry cut.

6   Symmetry Advantage in Tree Search Algorithms

Now we will show how these detected symmetrical literals can be used to increase the efficiency of SAT-based algorithms for the itemset mining. We choose in our implementation the Davis Putnam (DP) procedure to be the baseline method that we want to improve by the advantage of local symmetry elimination. We will show in the next session how the symmetry cut had been integrated in a DPLL solver.

If I is an inconsistent partial interpretation in which the assignment of the value $True$ to the current literal ℓ is shown to be conflicting, then according to Corollary 1, all the literals in the orbit of ℓ computed by using the group $Sym(L_{CNF(k,\mathcal{D})_I})$ returned by Saucy are symmetrical to ℓ. Thus, we assign the value $False$ to each literal in $\ell^{Sym(L_{CNF(k,\mathcal{D})})}$ since the value $True$ is shown to be contradictory, and then we prune the sub-space which corresponds to the value $True$ assignments. The resulting procedure called Satisfiable is given in Algorithm 1.

Procedure Satisfiable($\mathcal{F}$);
begin
    if $\mathcal{F} = \emptyset$ **then** $\mathcal{F}$ is satisfiable
    else if $\mathcal{F}$ contains the empty clause, **then** $\mathcal{F}$ is unsatisfiable
       else begin
          if there exists a mono-literal or a monotone literal ℓ **then**
             if Satisfiable($\mathcal{F}_\ell$) **then** $\mathcal{F}$ is satisfiable
             else $\mathcal{F}$ is unsatisfiable
          else begin
             Choose an unsigned literal ℓ of $\mathcal{F}$
             if Satisfiable($\mathcal{F}_\ell$) **then** $\mathcal{F}$ is satisfiable
             else
             begin
                Gen=Saucy($\mathcal{F}$);
                $\ell^{Sym(L_{\mathcal{F}})}$=orbit($\ell$,$Gen$)={$\ell_1, \ell_2, ..., \ell_n$};
                if Satisfiable($\mathcal{F}_{\neg\ell_1 \wedge \neg\ell_2 \wedge ... \wedge \neg\ell_n}$) **then** $\mathcal{F}$ is satisfiable
                else $\mathcal{F}$ is unsatisfiable
           end
          end
end

Fig. 1. The Davis Putnam procedure with local symmetry elimination

The input formula $\mathcal{F}$ expresses the boolean encoding $CNF(k, \mathcal{D})$. The function $orbit(\ell, Gen)$ is elementary, it computes the orbit of the literal ℓ from the set of generators Gen returned by Saucy.

7   Experiments

Now we shall investigate the performances of our search techniques by experimental analysis.

7.1   The Input Data-sets

We choose for our experiments the following data-sets:

- **Simulated data-sets:** In this class, we use the simulated data-sets, generated specifically to involve interesting symmetries. The data are available at http://www.cril.fr/decMining.
- **Public datasets:** The datasets used in this class are well known in the data mining community and are available at https://dtai.cs.kuleuven.be/CP4IM/datasets/

7.2   The Experimented Methods

Now we shall investigate the performances of our search techniques by experimental analysis. We choose the previous datasets for our study to show the symmetry behavior in solving the itemset mining problem. We expect that symmetry breaking will be profitable in other datasets. Here, we tested and compared three methods:

1. **No-sym:** search without symmetry breaking by using the AVAL solver [7] as the baseline method;
2. **Gl-sym** search with global symmetry breaking. This method uses in pre-processing phase the program SHATTER [3, 4] that detects and eliminates the global symmetries of the considered instance by adding on it symmetry breaking clauses, then apply the solver AVAL [7] to the resulting instance. The CPU time of *Gl-sym* includes the time that SHATTER spends to compute the global symmetry.
3. **Lo-sym:** search with local symmetry breaking. This method implements in AVAL the dynamic local symmetry detection and elimination strategy described in this work. The CPU time of *Lo-sym* includes local symmetry search time.

The common baseline search method for the three previous methods is AVAL. The complexity indicators are the CPU time and the size of the output. Both the time needed for computing local symmetry and global symmetry are added to the total CPU time of search. The source codes are written in C and compiled on a Core2Duo E8400, 2.8 GHZ and 4 Gb of RAM.

7.3   The Obtained Results

We reported in Figure 2 the practical results of the methods: *No-sym*, *Gl-sym*, and *Lo-sym*, on a simulated data *dataset-gen-jss-5* for the closed frequent itemset mining problem. The curves give the CPU times (the ones on the left in the figure) respectively the number of patterns (the ones on the right in the figure) with respect to the minimum support threshold. We can see on the time curves that symmetry elimination is profitable for the itemset mining problem. Indeed, both *Gl-sym* and *Lo-sym* outperform *No-sym*. We also remark that *Lo-sym* detects and eliminates more symmetries than *Gl-sym* and is more efficient. From the curves giving the number of patterns we can see that symmetry leads to significantly decrease the size of the output by keeping only non-symmetrical patterns. We can see that *Lo-sym* reduces more the output than *Gl-sym*. Local symmetry elimination is profitable for solving the itemset mining problem and outperforms dramatically global symmetry breaking on these problems.

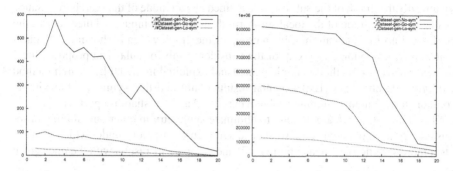

Fig. 2. Results on simulated data (Closed frequent itemsets): CPU time and number of patterns

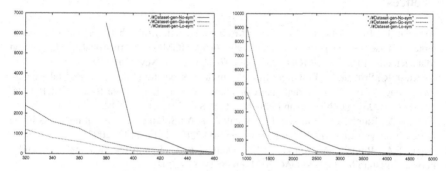

Fig. 3. Results on public data - *Australian and Muchroom* - (frequent itemsets): CPU time

In Figure 3, we reported the practical results of the methods *No-sym* and *Gl-sym* and *Lo-sym* on some public datasets for the frequent itemset mining problem. We can see that there exist some symmetries that are exploited and even the symmetries do not abound, *Gl-sym* and *Lo-sym* outperforms *No-sym* in CPU time. Indeed, many symmetrical no-good branches in the search tree are avoided in the exploration. However, the outputs are not reduced, since the detected symmetries usually involve items of the same transactions. We expect to reduce the size of the output when the detected symmetries involve items of different transactions.

8   Conclusion

We studied in this work the notions of global and local symmetry for the itemset mining problem expressed as a CNF formulas. We addressed the problem of dynamic symmetry detection and elimination of local symmetry during the search process. That is, the symmetries of each CNF sub-formula defined at a given node of the search tree and which is derived from the initial formula by considering the partial assignment corresponding to that node. Saucy is adapted to compute this local symmetry by maintaining dynamically the graph of the sub-formula defined at each node of the search tree. Saucy is called with the graph of the local sub-formula as the main input, and then returns the set of generators of the automorphism group of the graph which is shown to be equivalent to the local symmetry group of the considered sub-formula. The proposed local symmetry detection method is implemented and exploited in the DPLL search method to improve its efficiency. Experimental results confirmed that symmetry breaking is profitable for the itemset mining problem expressed as a satisfiability problem.

As a future work, we are looking to eliminate symmetry in other data mining problems and try to implement some weakened symmetry conditions under which we may detect more symmetries, then experiment it and compare its results with the ones given here.

References

1. Agrawal, R., Imieliński, T., Swami, A.: Mining association rules between sets of items in large databases. In: Proceedings of the 1993 ACM SIGMOD International Conference on Management of Data, SIGMOD 1993, pp. 207–216. ACM, New York (1993)
2. Agrawal, R., Srikant, R.: Fast algorithms for mining association rules in large databases. In: Proceedings of the 20th International Conference on Very Large Data Bases, VLDB 1994, pp. 487–499. Morgan Kaufmann Publishers Inc., San Francisco (1994)
3. Aloul, F.A., Ramani, A., Markov, I.L., Sakallak, K.A.: Solving difficult sat instances in the presence of symmetry. IEEE Transaction on CAD 22(9), 1117–1137 (2003)
4. Aloul, F.A., Ramani, A., Markov, I.L., Sakallak, K.A.: Symmetry breaking for pseudo-boolean satisfiabilty. In: ASPDAC 2004, pp. 884–887 (2004)
5. Aloul, F.A., Ramani, A., Markov, I.L., Sakallah, K.A.: Solving difficult SAT instances in the presence of symmetry. In: Proceedings of the 39th Design Automation Conference (DAC 2002), pp. 731–736. ACM Press (2002)
6. Aloul, F.A., Ramani, A., Markov, I.L., Sakallah, K.A.: Solving difficult instances of boolean satisfiability in the presence of symmetry. IEEE Trans. on CAD of Integrated Circuits and Systems 22(9), 1117–1137 (2003)
7. Audemard, G., Benhamou, B., Siegel, P.: AVAL: An enumerative method for SAT. In: Palamidessi, C., Moniz Pereira, L., Lloyd, J.W., Dahl, V., Furbach, U., Kerber, M., Lau, K.-K., Sagiv, Y., Stuckey, P.J. (eds.) CL 2000. LNCS (LNAI), vol. 1861, pp. 373–383. Springer, Heidelberg (2000)
8. Benhamou, B., Sais, L.: Theoretical study of symmetries in propositional calculus and application. In: Kapur, D. (ed.) CADE 1992. LNCS, vol. 607, pp. 281–294. Springer, Heidelberg (1992)
9. Benhamou, B., Sais, L.: Tractability through symmetries in propositional calculus. JAR 12, 89–102 (1994)
10. Benhamou, B., Saïdi, M.R.: Local symmetry breaking during search in csps. In: Bessière, C. (ed.) CP 2007. LNCS, vol. 4741, pp. 195–209. Springer, Heidelberg (2007)
11. Benhamou, B., Sais, L.: Theoretical study of symmetries in propositional calculus and applications. In: Kapur, D. (ed.) CADE 1992. LNCS, vol. 607, pp. 281–294. Springer, Heidelberg (1992)
12. Benhamou, B., Sais, L.: Tractability through symmetries in propositional calculus. J. Autom. Reasoning 12(1), 89–102 (1994)
13. Burdick, D., Calimlim, M., Gehrke, J.: Mafia: A maximal frequent itemset algorithm for transactional databases. In: ICDE, pp. 443–452 (2001)

14. Crawford, J., Ginsberg, M., Luks, E., Roy, A.: Symmetry-breaking predicates for search problems. In: Knowledge Representation (KR), pp. 148–159. Morgan Kaufmann (1996)
15. Desrosiers, C., Galinier, P., Hansen, P., Hertz, A.: Improving frequent subgraph mining in the presence of symmetry. In: MLG (2007)
16. Grahne, G., Zhu, J.: Fast algorithms for frequent itemset mining using fp-trees. IEEE Trans. on Knowl. and Data Eng. 17(10), 1347–1362 (2005)
17. Guns, T., Dries, A., Tack, G., Nijssen, S., Raedt, L.D.: Miningzinc: A modeling language for constraint-based mining. In: International Joint Conference on Artificial Intelligence, Beijing, China (August 2013)
18. Guns, T., Nijssen, S., De Raedt, L.: Itemset mining: A constraint programming perspective. Artif. Intell. 175(12-13), 1951–1983 (2011)
19. Gély, A., Medina, R., Nourine, L., Renaud, Y.: Uncovering and reducing hidden combinatorics in guigues-duquenne bases. In: Ganter, B., Godin, R. (eds.) ICFCA 2005. LNCS (LNAI), vol. 3403, pp. 235–248. Springer, Heidelberg (2005)
20. Han, J., Pei, J., Yin, Y.: Mining frequent patterns without candidate generation. In: Proceedings of the 2000 ACM SIGMOD International Conference on Management of Data, SIGMOD 2000, pp. 1–12. ACM, New York (2000)
21. Henriques, R., Lynce, I., Manquinho, V.M.: On when and how to use sat to mine frequent itemsets. CoRR abs/1207.6253 (2012)
22. Jabbour, S., Khiari, M., Sais, L., Salhi, Y., Tabia, K.: Symmetry-based pruning in itemset mining. In: 25th International Conference on Tools with Artificial Intelligence(ICTAI 2013). IEEE Computer Society, Washington DC (2013)
23. Jabbour, S., Sais, L., Salhi, Y.: Boolean satisfiability for sequence mining. In: CIKM, pp. 649–658 (2013)
24. Jabbour, S., Sais, L., Salhi, Y.: The top-k frequent closed itemset mining using top-k SAT problem. In: Blockeel, H., Kersting, K., Nijssen, S., Železný, F. (eds.) ECML PKDD 2013, Part III. LNCS, vol. 8190, pp. 403–418. Springer, Heidelberg (2013)
25. Jabbour, S., Sais, L., Salhi, Y., Tabia, K.: Symmetries in itemset mining. In: 20th European Conference on Artificial Intelligence (ECAI 2012), pp. 432–437. IOS Press (August 2012)
26. Khiari, M., Boizumault, P., Crémilleux, B.: Constraint programming for mining n-ary patterns. In: Cohen, D. (ed.) CP 2010. LNCS, vol. 6308, pp. 552–567. Springer, Heidelberg (2010)
27. Krishnamurthy, B.: Short proofs for tricky formulas. Acta Inf. 22(3), 253–275 (1985)
28. Krishnamurty, B.: Short proofs for tricky formulas. Acta Inf. (22), 253–275 (1985)
29. Métivier, J.P., Boizumault, P., Crémilleux, B., Khiari, M., Loudni, S.: A constraint language for declarative pattern discovery. In: Proceedings of the 27th Annual ACM Symposium on Applied Computing, SAC 2012, pp. 119–125. ACM, New York (2012)
30. Minato, S.I.: Symmetric item set mining based on zero-suppressed bDDs. In: Todorovski, L., Lavrač, N., Jantke, K.P. (eds.) DS 2006. LNCS (LNAI), vol. 4265, pp. 321–326. Springer, Heidelberg (2006)
31. Minato, S.I., Uno, T., Arimura, H.: Fast generation of very large-scale frequent itemsets using a compact graph-based representation (2007)
32. Murtagh, F., Contreras, P.: Hierarchical clustering for finding symmetries and other patterns in massive, high dimensional datasets. CoRR abs/1005.2638 (2010)
33. Raedt, L.D., Guns, T., Nijssen, S.: Constraint programming for itemset mining. In: KDD, pp. 204–212 (2008)
34. Raedt, L.D., Guns, T., Nijssen, S.: Constraint programming for data mining and machine learning. In: AAAI (2010)
35. Tseitin, G.S.: On the complexity of derivation in propositional calculus. In: Structures in the Constructive Mathematics and Mathematical Logic, pp. 115–125. H.A.O Shsenko (1968)

36. Uno, T., Asai, T., Uchida, Y., Arimura, H.: Lcm: An efficient algorithm for enumerating frequent closed item sets. In: Proceedings of Workshop on Frequent itemset Mining Implementations, FIMI 2003 (2003)
37. Uno, T., Kiyomi, M., Arimura, H.: Lcm ver. 2: Efficient mining algorithms for frequent/closed/maximal itemsets. In: FIMI (2004)
38. Vanetik, N.: Mining graphs with constraints on symmetry and diameter. In: Shen, H.T., Pei, J., Özsu, M.T., Zou, L., Lu, J., Ling, T.-W., Yu, G., Zhuang, Y., Shao, J. (eds.) WAIM 2010. LNCS, vol. 6185, pp. 1–12. Springer, Heidelberg (2010)
39. Zaki, M.J., Hsiao, C.-J.: Efficient algorithms for mining closed itemsets and their lattice structure. IEEE Trans. on Knowl. and Data Eng. 17(4), 462–478 (2005)

A Distance-Based Decision in the Credal Level

Amira Essaid[1,2], Arnaud Martin[2], Grégory Smits[2],
and Boutheina Ben Yaghlane[3]

[1] LARODEC, University of Tunis, ISG Tunis, Tunisia
[2]IRISA, University of Rennes1, Lannion, France
[3] LARODEC, University of Carthage, IHEC Carthage, Tunisia

Abstract. Belief function theory provides a flexible way to combine information provided by different sources. This combination is usually followed by a decision making which can be handled by a range of decision rules. Some rules help to choose the most likely hypothesis. Others allow that a decision is made on a set of hypotheses. In [6], we proposed a decision rule based on a distance measure. First, in this paper, we aim to demonstrate that our proposed decision rule is a particular case of the rule proposed in [4]. Second, we give experiments showing that our rule is able to decide on a set of hypotheses. Some experiments are handled on a set of mass functions generated randomly, others on real databases.

Keywords: belief function theory, imprecise decision, distance.

1 Introduction

Belief function theory [2,9] allows us to represent all kinds of ignorance and offers rules for combining several imperfect information provided by different sources in order to get a more coherent one. The combination process helps to make decisions later. Decision making consists in selecting, for a given problem, the most suitable actions to take. Today, we are often confronted with the challenge of making decisions in cases where information is imprecise or even not available. In [12], Smets proposed the transferable belief model (TBM) as an interpretation of the theory of belief functions. The TBM emphasizes a distinction between knowledge modeling and decision making. Accordingly, we distinguish the credal level and the pignistic level. In the credal level, knowledge is represented as belief functions and then combined. The pignistic level corresponds to decision making, a stage in which belief functions are transformed into probability functions.

The pignistic probability, the maximum of credibility and the maximum of plausibility are rules that allow a decision on a singleton of the frame of discernment. Sometimes and depending on application domains, it seems to be more convenient to decide on composite hypotheses rather than a simple one. In the literature, there are few works that propose a rule or an approach for making decision on a union of hypotheses [4,1,8]. Recently, we proposed a decision rule based on a distance measure [6]. This rule calculates the distance between a combined mass function and a categorical one. The most likely hypothesis to

G.A. Aranda-Corral et al. (Eds.): AISC 2014, LNAI 8884, pp. 147–156, 2014.

choose is the hypothesis whose categorical mass function is the nearest to the combined one.

The main topic of this paper is to demonstrate that our proposed decision rule is a particular case of that detailed in [4] and to extend our rule so that it becomes able to give decisions even with no categorical mass functions. We present also our experiments on mass functions generated randomly as well as on real databases.

The remainder of this paper is organized as follows: in section 2 we recall the basic concepts of belief function theory. Section 3 presents our decision rule based on a distance measure proposed in [6]. In section 4, we demonstrate that our proposed rule is a particular case of that proposed in [4]. Section 5 presents experiments and the main results. Section 6 concludes the paper.

2   The Theory of Belief Functions

The theory of belief functions [2,9] is a general mathematical framework for representing beliefs and reasoning under uncertainty. In this section, we recall some concepts of this theory.

The *frame of discernment* $\Theta = \{\theta_1, \theta_2, \ldots, \theta_n\}$ is a set of n elementary hypotheses related to a given problem. These hypotheses are exhaustive and mutually exclusive. The power set of Θ, denoted by 2^{Θ} is the set containing singleton hypotheses of Θ, all the disjunctions of these hypotheses as well as the empty set.

The Basic belief assignment *(bba)*, denoted by m is a mass function defined on 2^{Θ}. It affects a value from $[0, 1]$ to each subset. It is defined as:

$$\sum_{A \subseteq 2^{\Theta}} m(A) = 1. \tag{1}$$

A focal element A is an element of 2^{Θ} such that $m(A) > 0$. A categorical bba is a bba with a unique focal element such that $m(A) = 1$. When this focal element is a disjunction of hypotheses then the bba models imprecision.

Based on the basic belief assignment, other belief functions (credibility function ad plausibility function) can be deduced.

- Credibility function *bel(A)* expresses the total belief that one allocates to A. It is a mapping from elements of 2^{Θ} to $[0, 1]$ such that:

$$bel(A) = \sum_{B \subseteq A, B \neq \emptyset} m(B). \tag{2}$$

- Plausibility function *pl(A)* is defined as:

$$pl(A) = \sum_{A \cap B \neq \emptyset} m(B). \tag{3}$$

The plausibility function measures the maximum amount of belief that supports the proposition A by taking into account all the elements that do not contradict. The value $pl(A)$ quantifies the maximum amount of belief that might support a subset A of Θ.

The theory of belief function is a useful tool for data fusion. In fact, for a given problem and for the same frame of discernment, it is possible to get a mass function synthesizing knowledge from separate and independent sources of information through applying a combination rule. Mainly, there exists three modes of combination:

- Conjunctive combination is used when two sources are distinct and fully reliable. In [10], the author proposed the conjunctive combination rule which is defined as:

$$m_{1\textcircled{$\cap$}2}(A) = \sum_{B\cap C=A} m_1(B) \times m_2(C). \tag{4}$$

The Dempster's rule of combination [2] is a normalized form of the rule described previously and is defined as:

$$m_{1\oplus 2}(A) = \begin{cases} \dfrac{\displaystyle\sum_{B\cap C=A} m_1(B) \times m_2(C)}{1-\displaystyle\sum_{B\cap C=\emptyset} m_1(B) \times m_2(C)} & \forall A \subseteq \Theta,\ A \neq \emptyset \\ 0 & \text{if } A = \emptyset \end{cases} \tag{5}$$

This rule is normalized through $1-\displaystyle\sum_{B\cap C=\emptyset} m_1(B) \times m_2(C)$ and it works under the closed world assumption where all the possible hypotheses of the studied problem are supposed to be enumerated on Θ.

- Disjunctive combination: In [11], Smets introduced the disjunctive combination rule which combines mass functions when an unknown source is unreliable. This rule is defined as:

$$m_{1\textcircled{$\cup$}2}(A) = \sum_{B\cup C=A} m_1(B) \times m_2(C) \tag{6}$$

- Mixed combination: In [5], the authors proposed a compromise in order to consider the benefits of the two combination modes previously described. This combination is given for every $A \in 2^{\Theta}$ by the following formula:

$$\begin{cases} m_{DP}(A) = m_{1\textcircled{$\cap$}}(A) + \displaystyle\sum_{B\cap C=\emptyset, B\cup C=A} m_1(B)m_2(C) \ \forall A \in 2^{\Theta},\ A \neq \emptyset \\ m_{DP}(\emptyset) = 0 \end{cases}$$

$$\tag{7}$$

3   Decision Making in the Theory of Belief Functions

In the transferable belief model, decision is made on the pignistic level where the belief functions are transformed into a probability function, named *pignistic probability*. This latter, noted as *BetP* is defined for each $X \in 2^\Theta$, $X \neq 0$ as:

$$betP(X) = \sum_{Y \in 2^\Theta, Y \neq \emptyset} \frac{|X \cap Y|}{|Y|} \frac{m(Y)}{1 - m(\emptyset)} \tag{8}$$

where $|Y|$ represents the cardinality of Y.

Based on the obtained pignistic probability, we select the most suitable hypothesis with the maximum *BetP*. This decision results from applying tools of decision theory [4]. In fact, if we consider an entity represented by a feature vector x. A is a finite set of possible actions $A = \{a_1, \ldots, a_N\}$ and Θ a finite set of hypotheses, $\Theta = \{\theta_1, \ldots, \theta_M\}$. An action a_j corresponds to the action of choosing the hypothesis θ_j. But, if we select a_i as an action whereas the hypothesis to be considered is rather θ_j then the loss occurred is $\lambda(a_i|\theta_j)$. The expected loss associated with the choice of the action a_i is defined as:

$$R_{betP}(a_i|x) = \sum_{\theta_j \in \Theta} \lambda(a_i|\theta_j) BetP(\theta_j). \tag{9}$$

Then, the decision consists in selecting the action which minimizes the expected loss. In addition to minimizing pignistic expected loss, other risks are presented in [4].

Decision can be made on composite hypotheses [1,8]. We present in this paper the Appriou's rule [1] which helps to choose a solution of a given problem by considering all the elements contained in 2^Θ. This approach weights the decision functions (maximum of credibility, maximum of plausibility and maximum of pignistic probability) by an utility function depending on the cardinality of the elements. $A \in 2^\Theta$ is chosen if:

$$A = \underset{X \in 2^\Theta}{argmax}(m_d(X)pl(X)) \tag{10}$$

where m_d is a mass defined by:

$$m_d(X) = K_d \lambda_X \left(\frac{1}{|X|^r} \right) \tag{11}$$

The value r is a parameter in $[0, 1]$ helping to choose a decision which varies from a total indecision when r is equal to 0 and a decision based on a singleton when r is equal 1. λ_X helps to integrate the lack of knowledge about one of the elements of 2^Θ. K_d is a normalization factor and $pl(X)$ is a plausibility function.

In the following, we present our decision rule based on a distance measure.

4   Decision Rule Based on a Distance Measure

In [6], we proposed a decision rule based on a distance measure. It is defined as:

$$A = argmin(d(m_{comb}, m_A)) \tag{12}$$

This rule aims at deciding on a union of singletons. It is based on the use of categorical bba which helps to adjust the degree of imprecision that has to be kept when deciding. Depending on cases, we can decide on unions of two elements or three elements, etc. The rule calculates the distance between a combined bba m_{comb} and a categorical one m_A. The minimum distance is kept and the decision corresponds to the categorical bba's element having the lowest distance with the combined bba. The rule is applied as follows:

- We consider the elements of 2^Θ. In some applications, 2^Θ can be of a large cardinality. For this reason, we may choose some elements to work on. For example, we can keep the elements of 2^Θ whose cardinality is less or equal to 2.
- For each selected element, we construct its corresponding categorical bba.
- Finally, we apply Jousselme distance [7] to calculate the distance between the combined bba and a categorical bba. The distance with the minimum value is kept. The most likely hypothesis to select is the hypothesis whose categorical bba is the nearest to the combined bba.

Jousselme distance is defined for two bbas m_1 and m_2 as follows:

$$d(m_1, m_2) = \sqrt{\frac{1}{2}(m_1 - m_2)^t \underline{\underline{D}}(m_1 - m_2)} \tag{13}$$

where $\underline{\underline{D}}$ is a matrix based on Jaccard distance as a similarity measure between focal elements. This matrix is defined as:

$$D(A, B) = \begin{cases} 1 & \text{if A=B=}\emptyset \\ \frac{|A \cap B|}{|A \cup B|} & \forall A, B \in 2^\Theta \end{cases} \tag{14}$$

In this paper, we propose to apply the rule through two different manners:

- *Distance type 1* is calculated with categorical bbas ($m(A) = 1$) for all elements of 2^Θ except Θ to have an imprecise result rather than a total ignorance.
- *Distance type 2* is calculated with simple bbas such as $m(A) = \alpha$, $m(\Theta) = 1 - \alpha$.

In the following, we show that our proposed rule can be seen as a particular case of that proposed in section 3.

Jousselme distance can be written as:

$$d(m_1, m_2) = \frac{1}{2} \sum_{Y \subseteq \Theta} \sum_{X \subseteq \Theta} \frac{|X \cap Y|}{|X \cup Y|} m(X) m(Y) \tag{15}$$

If we consider the expected loss of choosing a_i, then it can be written as:

$$R_{betP}(a_i|x) = \sum_{Y \in \Theta} \lambda(a_i|Y) BetP(Y).$$

$$R_{betP}(a_i|x) = \sum_{Y \in \Theta} \lambda(a_i|Y) \sum_{X \in \Theta} \frac{|X \cap Y|}{|X|} \frac{m(X)}{1 - m(\emptyset)}. \qquad (16)$$

$$R_{betP}(a_i|x) = \sum_{Y \in \Theta} \sum_{X \in \Theta} \lambda(a_i|Y) \frac{|X \cap Y|}{|X|} \frac{m(X)}{1 - m(\emptyset)}.$$

The equation relative to decision is equal to that for the risk for a value of λ that has to be equal to:

$$\lambda(a_i|Y) = \frac{|X|(1 - m(\emptyset))}{|X \cup Y|} m(X) \qquad (17)$$

In this section, we showed that for a particular value of λ, our proposed decision rule can be considered as a particular case of that proposed in [4]. In the following section, we give experiments and present comparisons between our decision rule based on a distance measure and that presented in [1].

5   Experiments

5.1   Experiments on Generated Mass Functions

We tested the proposed rule [6] on a set of mass functions generated randomly. To generate the bbas, one needs to specify the cardinality of the frame of discernment, the number of mass functions to be generated as well as the number of focal elements. The generated bbas are then combined. We use the Dempster's rule of combination, the disjunctive rule and the mixed rule. Suppose we have a frame of discernment represented as $\Theta = \{\theta_1, \theta_2, \theta_3\}$ and three different sources for which we generate their corresponding bbas as given in Table 1.

Table 1. Three sources with their bbas

	S_1	S_2	S_3
θ_1	0.410	0.223	0.034
θ_2	0.006	0.108	0.300
$\theta_1 \cup \theta_2$	0.039	0.027	0.057
θ_3	0.026	0.093	0.128
$\theta_1 \cup \theta_3$	0.094	0.062	0.04
$\theta_2 \cup \theta_3$	0.199	0.153	0.004
$\theta_1 \cup \theta_2 \cup \theta_3$	0.226	0.334	0.437

We apply combination rules and we get the results illustrated in Table 2.

Table 2. Combination results

	Dempster rule	Disjunctive rule	Mixed rule
θ_1	0.369	0.003	0.208
θ_2	0.227	0	0.128
$\theta_1 \cup \theta_2$	0.025	0.061	0.075
θ_3	0.168	0	0.094
$\theta_1 \cup \theta_3$	0.049	0.037	0.064
$\theta_2 \cup \theta_3$	0.103	0.035	0.093
$\theta_1 \cup \theta_2 \cup \theta_3$	0.059	0.864	0.338

Table 3. Decision results

	Pignistic Probability	Appriou rule	Rule based on distance measure
Dempster rule	θ_1	$\theta_1 \cup \theta_2$	θ_1
Disjunctive rule	θ_1	θ_1	$\theta_1 \cup \theta_2$
Mixed rule	θ_1	θ_1	$\theta_1 \cup \theta_2$

Once the combination is performed, we can make decision. In Table 3, we compare between the results of three decision rules, namely the pignistic probability, the Appriou's rule with r equal to 0.5 as well as our proposed decision rule based on distance measure.

Table 3 shows the decision results obtained after applying some combination rules. We depict from this table that not all the time the rule proposed by Appriou gives a decision on a composite hypotheses. In fact, as shown in Table 3, the application of disjunctive rule as well as the mixed rule lead to a decision on a singleton which is θ_1. This is completely different from what we obtain when we apply our proposed rule which promotes a decision on union of singletons when combining bbas. The obtained results seems to be convenient especially that the disjunctive and the mixed rules help to get results on unions of singletons.

5.2 Experiments on Real Databases

To test our proposed decision rule, we do some experiments on real databases (IRIS[1] and HaberMan's survival[2]). Iris is a dataset containing 150 instances, 4 attributes and 3 classes where each class refers to a type of iris plant. HaberMan is a dataset containing results study conducted at the University of Chicago's Billings Hospital on the survival of patients who had undergone surgery for breast cancer. This dataset contains 306 instances, 3 attributes and 2 classes

---

[1] http://archive.ics.uci.edu/ml/datasets/Iris
[2] http://archive.ics.uci.edu/ml/datasets/Haberman%27s+Survival

(1: patient survived 5 years or longer, 2: patient died within 5 years). For the classification, our experiments are handled in two different manners.

- First, we apply the k-NN classifier [3]. The results are illustrated in a confusion matrix as shown in Table 4 (left side).
- Second, we modify the k-NN classifier's algorithm based on the use of Dempster rule of combination, to make it able to combine belief functions through the mixed rule. Then, Appriou's rule and our proposed decision rule are applied to make decision. Results are illustrated in Table 4.

Table 4. Confusion Matrices for Iris

k-NN classifier			
	θ_1	θ_2	θ_3
θ_1	11	0	0
θ_2	0	11	2
θ_3	0	0	16

Appriou's rule						
	θ_1	θ_2	$\theta_1 \cup \theta_2$	θ_3	$\theta_1 \cup \theta_3$	$\theta_2 \cup \theta_3$
θ_1	11	0	0	0	0	0
θ_2	0	15	0	0	0	0
θ_3	0	1	0	13	0	0

Our decision rule						
	θ_1	θ_2	$\theta_1 \cup \theta_2$	θ_3	$\theta_1 \cup \theta_3$	$\theta_2 \cup \theta_3$
θ_1	10	0	0	0	0	0
θ_2	0	12	0	2	0	1
θ_3	0	0	0	13	0	2

The same tests are done for HaberMan's survival dataset. The results of applying k-NN classifier, Appriou's rule and our decision rule are given respectively in Table 5. For the classification of 40 sets chosen randomly from Iris, we remark that with the k-NN classifier, all the sets having θ_1 and θ_3 as corresponding classes are well classified and only two originally belonging to class θ_2 were classified as θ_3. Appriou's rule gives a good classification for sets originally belonging to classes θ_1 and θ_2 and thus promoting a result on singletons rather than on a union of singletons.

Considering the results obtained when applying our decision rule based on a distance type 1, we note that only 2 sets are not well classified and that 3 have $\theta_2 \cup \theta_3$ as a class. The obtained results are good because our method is based on an imprecise decision which is underlined by the fact of obtaining $\theta_2 \cup \theta_3$ as a class.

Considering HaberMan's survival dataset, we note that the k-NN classifier, Appriou's rule as well as our decision rule give the same results where among the sets originally belonging to θ_1, 34 are well classified and among the 18 belonging to θ_2, only 6 are well classified. We obtain the same results as the other rules because the HaberMan's survival dataset has only two classes and our method is based on getting imprecise decisions and excluding the ignorance.

All the experiments given previously are based on the use of distance type 1. The results shown below are based on distance type 2. In fact, we consider a

Table 5. Confusion Matrices for HaberMan's survival

k-NN classifier			Appriou's rule				Our decision rule			
	θ_1	θ_2		θ_1	θ_2	Θ		θ_1	θ_2	Θ
θ_1	34	4	θ_1	34	4	0	θ_1	34	4	0
θ_2	12	6	θ_2	12	6	0	θ_2	12	6	0

simple bba and each time, we assign a value α to an element of 2^Θ. The tested rule on Iris as illustrated in Table 6 (left side) gives better results with an $\alpha < 0.8$. In addition to that, we obtained decisions on a union of singletons. The tests done on HaberMan's survival as given in Table 6 (right side) shows that with $\alpha > 0.5$, we obtain a better rate of good classification although we did not obtain a good classification for the class θ_2 and no set belongs to Θ. We aim in the future to make experiments on other datasets because HaberMan's survival, for example, does only have 2 classes, so we do not have enough imprecise elements.

Table 6. Rates of good classification

	$\alpha < 0.8$	$\alpha >= 0.8$
Iris	0.95	0.675

	$\alpha <= 0.2$	$\alpha \in [0.3, 0.5]$	$\alpha > 0.5$
HaberMan's survival	0.786	0.803	0.821

6 Conclusion

In this paper, we presented a rule based on a distance measure. This decision rule helps to choose the most likely hypothesis based on the calculation of the distance between a combined bba and a categorical bba. The aim of the proposed decision rule is to give results on composite hypotheses. In this paper, we demonstrated that our proposed rule can be seen as a particular case of that proposed in [4]. We presented also the different experiments handled on generated mass functions as well as on real databases.

References

1. Appriou, A.: Approche générique de la gestion de l'incertain dans les processus de fusion multisenseur. Traitement du Signal 22, 307–319 (2005)
2. Dempster, A.P.: Upper and Lower probabilities induced by a multivalued mapping. Annals of Mathematical Statistics 38, 325–339 (1967)
3. Denoeux, T.: A k-nearest neighbor classification rule based on Dempster-Shafer Theory. IEEE Transactions on Systems, Man, and Cybernetics 25(5), 804–813 (1995)
4. Denoeux, T.: Analysis of evidence-theoric decision rules for pattern classification. Pattern Recognition 30(7), 1095–1107 (1997)
5. Dubois, D., Prade, H.: Representation and combination of uncertainty with belief functions and possibility measures. Computational Intelligence 4, 244–264 (1988)

6. Essaid, A., Martin, A., Smits, G., Ben Yaghlane, B.: Uncertainty in ontology matching: a decision rule based approach. In: Proceeding of the International Conference on Information Processing and Mangement Uncertainty, pp. 46–55 (2014)
7. Jousselme, A.L., Grenier, D., Bossé, E.: A New Distance Between Two Bodies of Evidence. Information Fusion 2, 91–101 (2001)
8. Martin, A., Quidu, I.: Decision support with belief functions theory for seabed characterization. In: Proceeding of the International Conference on Information Fusion, pp. 1–8 (2008)
9. Shafer, G.: A mathematical theory of evidence. Princeton University Press (1976)
10. Smets, P.: The Combination of Evidence in the Transferable Belief Model. IEEE Transactions on Pattern Analysis and Machine Intelligence 12(5), 447–458 (1990)
11. Smets, P.: Belief functions: The disjunctive rule of combination and the generalized Bayesian theorem. International Journal of Approximate Reasoning 9(1), 1–35 (1993)
12. Smets, P., Kennes, R.: The Transferable Belief Model. Artificial Intelligent 66, 191–234 (1994)
13. Yager, R.R.: On the Dempster-Shafer framework and new combination rules. Information Sciences 41, 93–137 (1987)

Multivalued Elementary Functions in Computer-Algebra Systems

David J. Jeffrey

Department of Applied Mathematics
University of Western Ontario
djeffrey@uwo.ca

Abstract. An implementation (in Maple) of the multivalued elementary inverse functions is described. The new approach addresses the difference between the single-valued inverse function defined by computer systems and the multivalued function which represents the multiple solutions of the defining equation. The implementation takes an idea from complex analysis, namely the *branch* of an inverse function, and defines an index for each branch. The branch index then becomes an additional argument to the (new) function. A benefit of the new approach is that it helps with the general problem of correctly simplifying expressions containing multivalued functions.

1 Introduction

The manner in which computer-algebra systems handle multivalued functions, specifically the elementary inverse functions, has been the subject of extensive discussions over many years. See, for example, [5,6,8]. The discussion has centred on the best way to handle possible simplifications, such as

$$\sqrt{z^2} = z \ ? \quad \arcsin(\sin z) = z \ ? \quad \ln(e^z) = z \ ? \tag{1}$$

In the 1980s, errors resulting from the incorrect application of these transformations were common. Since then, systems have improved and now they usually avoid simplification errors, although the price paid is often that no simplification is made when it could be. For example, MAPLE 18 fails to simplify

$$\sqrt{1-z}\sqrt{1+z} - \sqrt{1-z^2} \ ,$$

even though it is zero for all $z \in \mathbb{C}$, see [2,8]. Here a new way of looking at such problemsis presented.

The discussion of possible treatments has been made difficult by the many different interpretations placed on the same symbols by different groups of mathematicians. Sorting through these interpretations, and assessing which ones are practical for computer algebra systems, has been an extended process. In this paper, we shall not revisit in any detail the many past contributions to the discussion, but summarize them and jump to the point of view taken here.

G.A. Aranda-Corral et al. (Eds.): AISC 2014, LNAI 8884, pp. 157–167, 2014.
© Springer International Publishing Switzerland 2014

1.1   A Question of Values

One question which has been discussed at length concerns the number of values represented by function names. One influential point of view was expressed by Carathéodory, in his highly regarded book [4]. Considering the logarithm function, he addressed the equation

$$\ln z_1 z_2 = \ln z_1 + \ln z_2 , \tag{2}$$

for complex z_1, z_2. He commented [4, pp. 259–260]:

> The equation merely states that the sum of one of the (infinitely many) logarithms of z_1 and one of the (infinitely many) logarithms of z_2 can be found among the (infinitely many) logarithms of $z_1 z_2$, and conversely every logarithm of $z_1 z_2$ can be represented as a sum of this kind (with a suitable choice of $\ln z_1$ and $\ln z_2$).

In this statement, Carathéodory first sounds as though he thinks of $\ln z_1$ as a symbol standing for a set of values, but then for the purposes of forming an equation he prefers to select one value from the set. Whatever the exact mental image he had, the one point that is clear is that $\ln z_1$ does not have a unique value, which is in strong contrast to every computer system. Every computer system will accept a specific value for z_1 and return a unique $\ln z_1$.

The reference book edited by Abramowitz & Stegun [1, Chap 4] is another authoritative source, as is its successor [15]. They both define, to take one example, the solution of $\tan t = z$ to be $t = \text{Arctan}\, z = \arctan z + k\pi$. When listing properties, they both give the equation

$$\text{Arctan}(z_1) + \text{Arctan}(z_2) = \text{Arctan}\,\frac{z_1 + z_2}{1 - z_1 z_2} . \tag{3}$$

For $z_1 = z_2 = \sqrt{3}$, we have $\text{Arctan}\,\sqrt{3} + \text{Arctan}\,\sqrt{3} = \text{Arctan}(-\sqrt{3})$. For computer users, this is confusing, because their systems return values $\arctan\sqrt{3} = \pi/3$ and $\arctan(-\sqrt{3}) = -\pi/3$, and most users do not see the difference between Arctan and arctan. (Below, a new form of (3) is given.) By comparing the Abramowitz & Stegun definition with the statement of Carathéodory, we can see that as far as equations are concerned, both sets of authors favour an interpretation based on interactively selecting one value from a set of possible ones.

Riemann surfaces give a very pictorial way of seeing multi-valuedness [16,7], but a question remains whether they can be used computationally [13]. To discuss these approaches in detail will deflect attention from the implementation here. Therefore, now that alternative approaches have been noted, they will be set aside.

Here, an inverse function will have a single value [13]. Further, that single value will be determined by the arguments to the function and not by the context in which it finds itself.

2   A New Treatment of Inverse Functions

The basis of the new implementation is notation introduced in [11]. To the standard function $\ln z$, a subscript is added:

$$\ln_k z = \ln z + 2\pi i k \ .$$

Here the function $\ln z$ denotes the principal value of logarithm, which is the single-valued function with imaginary part $-\pi < \Im \ln z \leq \pi$. This is the function currently implemented in Maple, Mathematica, Matlab and other systems. In contrast, $\ln_k z$ denotes the kth branch of logarithm. With this notation, the statement above of Carathéodory can be restated unambiguously as

$$\exists k, m, n \in \mathbb{Z}, \text{ such that } \ln_k z_1 z_2 = \ln_m z_1 + \ln_n z_2 \ .$$

His "and conversely" statement is actually a stronger statement. He states

$$\forall k \in \mathbb{Z}, \exists m, n \in \mathbb{Z}, \text{ such that } \ln_k z_1 z_2 = \ln_m z_1 + \ln_n z_2 \ .$$

In the light of his converse statement, Carathéodory's first statement could be interpreted as meaning

$$\forall m, n \in \mathbb{Z}, \exists k \in \mathbb{Z}, \text{ such that } \ln_m z_1 + \ln_n z_2 = \ln_k z_1 z_2 \ .$$

I think the English statement does not support this interpretation, but it may be supported by the original German. In any event, it shows the greater conciseness of branch notation.

The principal of denoting explicitly the branch of a multivalued function will be extended here to all the elementary multivalued functions. In order for the new treatment to be smoothly implemented in Maple, a system of notation is needed that can co-exist with the built-in functions of Maple.

2.1   Notation for Inverses

The built-in functions for which we shall be implementing branched replacements are

- `log(z)`,
- `arcsin(z)`, `arccos(z)`, `arctan(z)`,
- `arcsinh(z)`, `arccosh(z)`, `arctanh(z)`,
- fractional powers $z^{1/n}$.

Rather than risk confusion by trying to modify the actions of these names within Maple, we shall leave the built-in functions untouched and work with independent, clearly defined and unambiguous notation for the branched functions.

The model we follow is to adapt the notation `invfunc` used in Maple; Mathematica has a similar construction `InverseFunction`. The most direct presentation is simply to display the definitions, with source code.

2.2   Subscripts in Maple

A subscript on a function f, as in $f_k(z)$, is really an additional argument to the function, except that instead of placing it in parentheses, as in $f(k, z)$, we choose subscripting. In Maple, however, the programming is quite different in the two cases. Thus $f(k, z)$ is coded as

`f := proc (k,z) ... end`

and the k and z can be used in the procedure without further programming. A subscripted function, however, is written as `f[k](z)`, and is an 'indexed name'. The procedure is now coded as

`f := proc (z) ... end`

and inside the procedure there is a variable available to the program called `procname`. If the procedure has been called with an indexed name, then this is contained in `procname` and the index, i.e., the subscript, can be retrieved for use in the procedure by using the `op` function.

3   Particular Functions

In this section, the inverses of the elementary functions are defined in the new notation. The implementations use Maple's indexed names, and in Maple's 2-D printing, the indexes appear as subscripts.

3.1   Inverse Sine

The principal branch of the inverse sine function is denoted in Maple by `arcsin`. Using this, we define the branched inverse sine by

$$\text{invsin}_0 z = \arcsin z \, , \tag{4}$$

$$\text{invsin}_k z = (-1)^k \, \text{invsin}_0 z + k\pi \, . \tag{5}$$

The principal branch now has the equivalent representation $\text{invsin}_0 z = \text{invsin} \, z = \arcsin z$. It has real part between $-\pi/2$ and $\pi/2$. Notice that the branches are spaced a distance π apart in accordance with the antiperiod[1] of sine, but the repeating unit is of length 2π in accord with the period of sine.

The Maple code for the function is

```
invsin := proc (z::algebraic) local branch;
            if nargs <> 1 then
                error "Expecting 1 argument, got", nargs ;
```

---

[1] An antiperiodic function is one for which $\exists \alpha$ such that $f(z + \alpha) = -f(z)$, and α is then the antiperiod. This is a special case of a quasi-periodic function [14], namely one for which $\exists \alpha, \beta$ such that $f(z + \alpha) = \beta f(z)$.

```
      elif type(procname, 'indexed') then
          branch := op(procname);
          branch*Pi+(-1)^branch*arcsin(z);
      else arcsin(z);
      end if;
  end proc;
```

The **nargs** function counts the number of arguments supplied by the user, and although here the code is restricted to 1 argument, one could allow the branch number to be passed as an argument instead of as a subscript. Note that the code is not 'industrial strength', and in particular the branch is not tested for being an integer. Since the code is exploratory, it relies on the user being sensible. Examples of its use appear below.

3.2   Inverse Cosine

The principal branch has real part between 0 and π, and this is easiest achieved by setting $\mathrm{invcos}_k z = \mathrm{invsin}_{k+1} z - \pi/2$. The code is

```
invcos := proc (z::algebraic) local branch;
          if nargs <> 1 then
              error "Expecting 1 argument, got", nargs ;
          elif type(procname, 'indexed') then
              branch := op(procname);
              invsin[branch+1](z)-Pi/2;
          else arccos(z);
          end if;
      end proc;
```

3.3   Inverse Tangent

The principal branch has real part from $-\pi/2$ to $\pi/2$, and the kth branch is $\mathrm{invtan}_k z = \mathrm{invtan}\, z + k\pi$. As code:

```
invtan := proc (z::algebraic) local branch;
          if nargs <> 1 then
              error "Expecting 1 argument, got", nargs ;
          elif type(procname, 'indexed') then
              branch := op(procname);
              branch*Pi+arctan(z);
          else arctan(z);
          end if;
      end proc;
```

The two-argument inverse tangent function has been implemented in many computer languages. It is a synonym for arg, in that $\arg(x + iy) = \arctan(y, x)$ for $x, y \in \mathbb{R}$. It can be described using the branches of invtan as

$$\arctan(y, x) = \mathrm{invtan}_k(y/x) \ ,$$

where $k = H(-x)\operatorname{sgn} y$, and H is the Heaviside step function. For x small this is inaccurate, when using invcot is better.

3.4   The Logarithm

The logarithm is the inverse of the exponential function, and therefore our convention would suggest implementing the branched version using `invexp`. This, however, seems too radical for acceptance, so we use `loge` instead. Another possibility might seem to be `Log`, but this is unsatisfactory because textbooks cannot agree on the definition of Log. Also Mathematica uses `Log[x]` as its standard log function, and may in the future have its own branch implementation.

```
loge := proc (z::algebraic) local branch;
           if nargs <> 1 then
                 error "Expecting 1 argument, got", nargs ;
           elif type(procname, 'indexed') then
                 branch := op(procname);
                 ln(z) + 2*Pi*I*branch;
           else ln(z);
           end if;
       end proc;
```

3.5   Inverse Hyperbolic Functions

A common point of contention in notation for inverse hyperbolic functions is whether to write `arcsinh` or `arsinh`, and similarly for the other functions. The point of the debate being that the geometrical interpretation of inverse sinh is an area, not an arc. Maple and Mathematica use the former notation to the chagrin of more enlightened authors [3,9] who prefer the latter. They argue that `arc` should not be merely a synonym for inverse. The convention here allows us to avoid this argument by using the `inv` prefix. We use the Russian abbreviations for the primary functions to save typing. Thus we define in the obvious way `invsh[k](z)`, `invch[k](z)`, `invth[k](z)`. We save space by not listing them.

3.6   Fractional Powers

The principal branch of $z^{1/n}$ is defined by $\exp(\frac{1}{n}\ln z)$, and replacing $\ln z$ by $\ln_k z$ gives the branched function. The standard notation for roots and fractional powers does not leave an obvious place for the branch label, and most obvious names are already used by Maple or Mathematica. We use the name `invpw`, meaning inverse (integer) power. The Maple code defines `invpw[k](z,n)`, where the subscript is the branch, as usual, while the fractional power is $1/n$. Thus it is modelled on the Maple `surd` function. Unlike the other inverse functions, there are only n distinct values, but we allow k to be any integer.

Since square root is so common, it is coded separately as `invsq[k](z)`, and it can be displayed in traditional notation as $(-1)^k\sqrt{z}$.

4 Applications

We now demonstrate some uses of the new notation.

4.1 Plotting

With the new functions, we can easily plot branches. Figure 1 shows plots produced by the Maple commands

```
> plot([invsin[-1](x),invsin(x),invsin[1](x)],x=-1 .. 1,
    linestyle=[2,1,3]);
> plot([invtan[-1](x),invtan(x),invtan[1](x),invtan[2](x)],
    x=-5..5, discont = true, linestyle = [2, 1, 3, 4]);
```

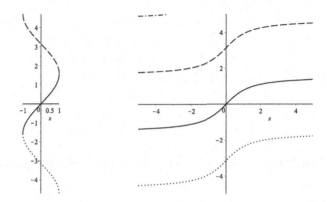

Fig. 1. The branches of inverse sine and inverse tangent plotted taking advantage of branch notation

4.2 Identities

In order to express identities containing inverse functions correctly, we need the unwinding number,

$$\mathcal{K}(z) = \left\lceil \frac{z - \pi}{2\pi} \right\rceil ,$$

defined in [5] (rather than in [6] where the sign is different). Note that the unwinding number is a built-in function in Maple, called unwindK. This immediately gives us

$$\ln_k e^z = z - 2\pi i \mathcal{K}(z) + 2\pi i k . \tag{6}$$

Note the special case $\ln_{\mathcal{K}(z)} e^z = z$.

Consider an identity one might see in a traditional treatment:

$$\cos x = \sqrt{1 - \sin^2 x} \ , \tag{7}$$

where the author would add "and the branch of the root is chosen appropriately". Using the branched root, we write the more precise

$$\cos x = \mathrm{invsq}[\mathcal{K}(2ix)](1 - \sin^2 x) = (-1)^{\mathcal{K}(2ix)} \sqrt{1 - \sin^2 x} \ . \tag{8}$$

We can contrast the two approaches in Maple with the command

```
> plot([ sqrt(1-sin(x)^2), invsq[unwindK(2*x*I)](1-sin(x)^2)],
     x = -7 .. 7, linestyle = [2, 1]);
```

The resulting plot is given in figure 2.

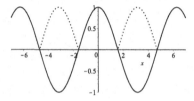

Fig. 2. The graph of $\sqrt{1 - \sin^2 x}$ using branch notation for square root

We return to the Abramowitz and Stegun [1] 'identity' (3). The branch problems with this equation are neatly displayed by the Maple command

```
> plot3d([arctan(x)+arctan(y), arctan((x+y)/(1-x*y))],
x = -2 .. 2, y = -2 .. 2, orientation = [-45, 45, 0])
```

The more precise identity is

$$\mathrm{invtan}(x) + \mathrm{invtan}(y) = \mathrm{invtan}_k \frac{x+y}{1-xy}, \ \text{ where } k = H(xy - 1)\,\mathrm{sgn}(x) \ , \tag{9}$$

and H is the Heaviside step. A more complicated example from [1] is their identity for $\mathrm{Arcsin}\,x + \mathrm{Arcsin}\,y$, which becomes

$$\mathrm{invsin}\,x + \mathrm{invsin}\,y = \mathrm{invsin}[k] \left(x\sqrt{1 - y^2} + y\sqrt{1 - x^2} \right) \ , \tag{10}$$

$$k = H(x^2 + y^2 - 1)(\mathrm{sgn}\,x + \mathrm{sgn}\,y)/2 \ .$$

Here the branch of invsin is allowed to vary, but there might be another formula which includes variable branches of square root.

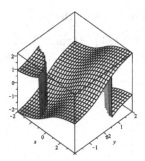

Fig. 3. A plot of the sum of two inverse tangents and the usual formula for their sum

As a final identity, we consider formula (4.4.39) in [1].

$$\text{Arctan}(x + iy) = k\pi + \frac{1}{2} \arctan \frac{2x}{1 - x^2 - y^2} + \frac{i}{4} \ln \frac{x^2 + (y + 1)^2}{x^2 + (y - 1)^2} .$$

To turn this identity into something that computer-algebra systems can use, one should decide what to do with k. This can be replaced by

$$\text{invtan}_k(x + iy) = \frac{1}{2} \text{invtan}_n \frac{2x}{1 - x^2 - y^2} + \frac{i}{4} \ln \frac{x^2 + (y + 1)^2}{x^2 + (y - 1)^2} ,$$

where $n = 2k + \text{sgn}(x)H(x^2 + y^2 - 1)$.

4.3 Calculus

Calculating the derivative of an inverse function is a standard topic in calculus. The results in the textbooks are restricted to the principal branches of the functions. It is possible, however, to generalize results to any branch. For example

$$\frac{d}{dx} \text{invsin}_k x = \frac{1}{\cos(\text{invsin}_k x)} = \frac{(-1)^k}{\sqrt{1 - x^2}} .$$

Integration by substitution is a well-known application of inverse functions. A specific difficulty has been the application of the substitution $u = \tan \frac{1}{2}x$ in integrals such as

$$\int \frac{3\,dx}{5 - 4\cos x} = \int \frac{6\,du}{1 + 9u^2} = 2\arctan(3\tan \tfrac{1}{2}x) . \tag{11}$$

The right-hand side is discontinuous, as has been pointed out in [12,10]. The correction to the usual integration formula [12] can be rewritten in the new notation as

$$\int \frac{3\,dx}{5 - 4\cos x} = 2\,\text{invtan}_{\mathcal{K}(ix)}(3\tan \tfrac{1}{2}x) . \tag{12}$$

The contrast is illustrated in figure 4 by the plot

```
> plot([ 2*invtan[unwindK(I*x)](3*tan((1/2)*x)),
      2*arctan(3*tan((1/2)*x))], x=-3..9,linestyle=[2,1],
        discont=true);
```

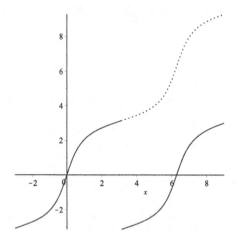

Fig. 4. A graph of the discontinuous and continuous integrals discussed in (11) and (12)

5   Conclusions

The focus here has been on the implementation of multivalued inverse functions in a computer-algebra system. The development of the notation, and of tools such as the unwinding number, has been motivated by the idea that traditional treatments of multivalued functions are not precise enough. Too often, decisions on branch choices are avoided by texts, the avoidance being often covered by phrases such as "taking an appropriate branch". The notation here allows one to state precisely which branch of a function should be used, and the notation also reminds one that such choices are important.

After branch information is added to existing equations, they are typically longer than before. This means that people looking for elegance rather than strictness will find little benefit in the new approach and notation. Looking back at Carathéodory's discussion of (2), we can see exactly the desire for elegance of presentation bringing with it the cost of impreciseness.

Outside computer-algebra systems, the notation also offers benefits. For example, it should make the topic easier for students. We already teach students that $y = x^2$ implies $x = \pm\sqrt{y}$, and we teach calculus students that $dy/dx = 1$ implies $y = x + K$, where K is a constant. So solutions to equations in which arbitrary elements appear are already part of a student's education. By using branch indexing, we can bring all the elementary inverse functions into a single pattern, and both students and computers are forced to confront branch choices explicitly.

There are many multivalued functions in mathematics, and here we have considered only the elementary functions. The principles developed here can be found already in Maple to varying degrees. The Lambert W function has been fully implemented using the same ideas of explicit branches as here. Maple's `RootOf` construction uses an index to specify different roots of an equation. Although there is a tendency to think of `RootOf` as specifying values rather than functions, there is no reason not to use it to define a function, although its generality will often make the branch structure of the defined function difficult to understand. The current approach is one of a number of possibilities for correct manipulation in a computer-algebra system. It fits together with the unwinding number approach happily and offers other ways of presenting and working with expressions. As with the unwinding number, there remains much scope for further development.

References

1. Abramowitz, M., Stegun, I.J.: Handbook of Mathematical Functions. Dover (1965)
2. Bradford, R.J., Corless, R.M., Davenport, J.H., Jeffrey, D.J., Watt, S.M.: Reasoning about the elementary functions of complex analysis. Annals of Mathematics and Artificial Intelligence 36, 303–318 (2002)
3. Bronshtein, I.N., Semendyayev, K.A., Musiol, G., Muehlig, H.: Handbook of Mathematics, 5th edn. Springer, Heidelberg (2007)
4. Carathéodory, C.: Theory of functions of a complex variable, 2nd edn. Chelsea, New York (1958)
5. Corless, R.M., Davenport, J.H., Jeffrey, D.J., Watt, S.M.: According to Abramowitz and Stegun. SIGSAM Bulletin 34, 58–65 (2000)
6. Corless, R.M., Jeffrey, D.J.: The unwinding number. Sigsam Bulletin 30(2), 28–35 (1996)
7. Corless, R.M., Jeffrey, D.J.: Elementary Riemann surfaces. Sigsam Bulletin 32(1), 11–17 (1998)
8. Davenport, J.H.: The challenges of multivalued "functions". In: Autexier, S., Calmet, J., Delahaye, D., Ion, P.D.F., Rideau, L., Rioboo, R., Sexton, A.P. (eds.) AISC 2010. LNCS, vol. 6167, pp. 1–12. Springer, Heidelberg (2010)
9. Gullberg, J.: Mathematics: From the Birth of Numbers. W. W. Norton & Company, New York (1997)
10. Jeffrey, D.J.: The importance of being continuous. Mathematics Magazine 67, 294–300 (1994)
11. Jeffrey, D.J., Hare, D.E.G., Corless, R.M.: Unwinding the branches of the Lambert W function. Mathematical Scientist 21, 1–7 (1996)
12. Jeffrey, D.J., Rich, A.D.: The evaluation of trigonometric integrals avoiding spurious discontinuities. ACM Trans. Math. Software 20, 124–135 (1994)
13. Jeffrey, D.J., Norman, A.C.: Not seeing the roots for the branches. SIGSAM Bulletin 38(3), 57–66 (2004)
14. Lawden, D.F.: Elliptic functions and applications. Springer (1989)
15. Lozier, D.W., Olver, F.W.J., Boisvert, R.F.: NIST Handbook of Mathematical Functions. Cambridge University Press (2010)
16. Trott, M.: Visualization of Riemann surfaces of algebraic functions. Mathematica in Education and Research 6, 15–36 (1997)

Rational Conchoid and Offset Constructions: Algorithms and Implementation

Juana Sendra[1,2], David Gómez[2], and Valerio Morán[3]

[1] Dpto. de Matemática Aplicada a las TIC, Universidad Politécnica de Madrid,
Madrid, Spain
jsendra@etsist.upm.es
[2] Research Center on Software Technologies and Multimedia Systems
for Sustainability, Madrid, Spain
david.gomezs@upm.es
[3] ETSI Sistemas de Telecomunicación, Universidad Politécnica de Madrid, Spain
vmoran@alumnos.upm.es

Abstract. This paper is framed within the problem of analyzing the rationality of the components of two classical geometric constructions, namely the offset and the conchoid to an algebraic plane curve and, in the affirmative case, the actual computation of parametrizations. We recall some of the basic definitions and main properties on offsets (see [13]), and conchoids (see [15]) as well as the algorithms for parametrizing their rational components (see [1] and [16], respectively). Moreover, we implement the basic ideas creating two packages in the computer algebra system Maple to analyze the rationality of conchoids and offset curves, as well as the corresponding help pages. In addition, we present a brief atlas where the offset and conchoids of several algebraic plane curves are obtained, their rationality analyzed, and parametrizations are provided using the created packages.

Keywords: Offset variety, conchoid variety, rational parametrization, symbolic mathematical software.

Introduction

In this paper we deal with two different geometric constructions that appear in many practical applications, where the need of proving rational parametrizations as well as automatized algorithmic processes is important. On one side we consider offset varieties and on the other conchoid varieties. Offsets varieties have been extensively applied in the field of computer aided geometric design (see [5],[3],[4]), while conchoids varieties appears in several of practical applications, namely the design of the construction of buildings, in astronomy [6], in electromagnetism research [20], optics, physics, mechanical engineering and biological engineering [7], in fluid mechanics [19], etc.

The intuitive idea of these geometric constructions is the following. Let $\mathbb{C}$ be the field of complex numbers (in general, one can take any algebraically closed field of characteristic zero), and let $\mathcal{C}$ be an irreducible hypersurface in $\mathbb{C}^n$ (say

G.A. Aranda-Corral et al. (Eds.): AISC 2014, LNAI 8884, pp. 168–179, 2014.

$n = 2$ or $n = 3$, and hence $\mathcal{C}$ is a curve or a surface). Moreover, although it is not necessary for the development of the theory, in practice one considers that $\mathcal{C}$ is real (i.e. there exists at least one regular real point on $\mathcal{C}$). *The offset variety* to $\mathcal{C}$ at distance d (d is a field element, in practice a non-zero real number), denoted by $\mathcal{O}_d(\mathcal{C})$, is the envelope of the system of hyperspheres centered at the points of $\mathcal{C}$ with fixed radius d (see Fig.1, left); for a formal definition, see e.g. [1]. In particular, if $\mathcal{C}$ is unirational and $\mathcal{P}(\bar{t})$, with $\bar{t} = (t_1, \ldots, t_n)$, a rational parametrization of $\mathcal{C}$, the offset to $\mathcal{C}$ is the Zariski closure of the set in $\mathbb{C}^n$ generated by the expression $\mathcal{P}(\bar{t}) \pm d \frac{\mathcal{N}(\bar{t})}{\|\mathcal{N}(\bar{t})\|}$ where $\mathcal{N}(\bar{t})$ is the normal vector to $\mathcal{C}$ associated with $\mathcal{P}(\bar{t})$.

The conchoid construction is also rather intuitive. Given $\mathcal{C}$ as above (base variety) and a fixed point A (focus), consider the line $\mathcal{L}$ joining A (in practice the focus is real) to a point P of $\mathcal{C}$. Now we take the points Q of intersection of $\mathcal{L}$ with a hypersphere of radius d centered at P. The Zariski closure of the geometric locus of Q as P moves along $\mathcal{C}$ is called *the conchoid variety* of $\mathcal{C}$ from focus A at a distance d and denoted by $\mathfrak{C}_d^A(\mathcal{C})$ (see Fig.1 right); for the geometric construction of the conchoid and, for a formal definition, see e.g. [15] and [10]. The *Conchoid of Nicomedes* and the *Limaçon of Pascal* are the two classic examples of conchoids, and the best known. They appear when the base curve is a line or a circle, respectively. Similarly, if $\mathcal{C}$ is unirational and $\mathcal{P}(\bar{t})$ is a rational parametrization of $\mathcal{C}$, then the conchoid is the Zariski closure of the set defined by the expression

$$\mathcal{P}(\bar{t}) \pm d \, \frac{\mathcal{P}(\bar{t}) - A}{\|\mathcal{P}(\bar{t}) - A\|} \, .$$

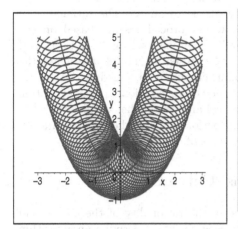
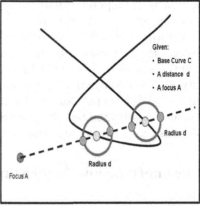

Fig. 1. Left: Construction of the offset to the parabola, Right: Geometric construction of the conchoid

These two operations are algebro-geometric, in the sense that they create new algebraic sets from the given input objects. There is an interesting relation between the offset and the conchoid operations. Indeed, there exists a rational bijective quadratic map which transforms a given hypersurface F and its offset F_d to a hypersurface G and its conchoidal G_d, and vice versa (see [11]).

The main difficulty when applying these constructions is that they generate much more complicated objects than the initial ones. There is a clear explosion of the degree of the hypersurface, singularity structure and the density of the defining polynomials (see e.g. [2], [13], [15]). As a consequence, in practice, the implicit equations are untractable from the computational point of view. This is one of the reasons why the use of parametric representations of offsets and conchoids are considered. Let us see an illustrating example. We consider the plane curve $\mathcal{C}$ defined by $y^4 = x^5$. Its offset has degree 12 and the polynomial defining it has 65 nonzero terms, and its infinity norm is 300781250. However, the offset can be parametrized by radicals (see [17], [18]) as $(t^4, t^5) \pm \frac{d}{\sqrt{25t^8 + 16t^6}}(-5t^4, 4t^3)$.
On the other, the offset can be rationally parametrized as

$$\left(\frac{1}{625} \frac{(t^2 - 1)\left(16\,t^8 - 32\,t^6 + 625\,dt^4 + 32\,t^2 - 16\right)}{t^4\left(t^2 + 1\right)}, \right.$$

$$\left. \frac{2}{3125} \frac{-16\,t^{12} + 64\,t^{10} - 80\,t^8 + 3125\,dt^6 + 80\,t^4 - 64\,t^2 + 16}{t^5\left(t^2 + 1\right)} \right) .$$

The paper is structured as follows. In Section 1 we recall some of the basic definitions and main properties on offsets and conchoids of algebraic plane curves (see [13], [15]). We provide algorithms to analyze the rationality of the components of these new objects (see [1], [16]), and in the affirmative case, rational parametrizations are given. In Section 2 we present the creation of two packages in the computer algebra system Maple to analyze the rationality of offset and conchoids curves respectively, whose procedures are based on the above algorithms, as well as the corresponding help pages. Finally, in Section 3, we illustrate the performance of the package by presenting a brief atlas where the offset and conchoids of several algebraic plane curves are obtained, with their rationality analyzed. Furthermore, in case of genus zero, a rational parametrization is computed. We have not done an theoretical analysis of the complexity of the implemented algorithms but the practical performance of the implementation provides answers, in reasonable time for curves of degree less than 5.

1   Parametrization Algorithms: Curve Case

In this section we summarize the results on the rationality of the offsets and conchoids of curves, presented in [1], [16] respectively, by deriving an algorithm for parametrizing them. The treatment of surfaces can be found at [1], [12], [14] (offsets) [8], [9], [10] (conchoids).

The Offset Rationality Problem

The rationality of the components of the offsets is characterized by means of the existence of parametrizations of the curve whose normal vector has rational norm, and alternatively by means of the rationality of the components of an associated curve, that is usually simpler than the offset. As a consequence, one deduces that offsets to rational curves behave as follows: they are either reducible with two rational components (double rationality), or rational, or irreducible and not rational.

For this purpose, we first introduce two concepts: Rational Pythagorean Hodograph and curve of reparametrization. Let $\mathcal{P}(t) = (P_1(t), P_2(t)) \in \mathbb{C}(t)^2$ be a rational parametrization of $\mathcal{C}$. Then, $\mathcal{P}(t)$ is RPH (*Rational Pythagorean Hodograph*) if its normal vector $\mathcal{N}(t) = (N_1(t), N_2(t))$ satisfies that $N_1(t)^2 + N_2(t)^2 = m(t)^2$, with $m(t) \in \mathbb{C}(t)$. For short we will express this fact writing $\|\mathcal{N}(t)\| \in \mathbb{C}(t)$. On the other hand, we define the *reparametrizing curve* of $\mathcal{O}_d(\mathcal{C})$ associated with $\mathcal{P}(t)$ as the curve generated by the primitive part with respect to x_2 of the numerator of $x_2^2 \, P_1'(x_1) - P_1'(x_1) + 2 \, x_2 \, P_2'(x_1)$, where P_i' denotes the derivative of P_i. In the following, we denote by $\mathcal{G}_{\mathcal{P}}^{\mathcal{O}}(\mathcal{C})$ the reparametrizing curve of $\mathcal{O}_d(\mathcal{C})$ associated with $\mathcal{P}(t)$. Summarizing the results in [1], one can outline the following algorithm for offsets.

Algorithm: offset parametrization

- GIVEN: a proper rational parametrization $\mathcal{P}(t)$ of a plane curve $\mathcal{C}$ in $\mathbb{K}^2$ and $d \in \mathbb{C}$.
- DECIDE: whether the components of $\mathcal{O}_d(\mathcal{C})$ are rational.
- DETERMINE: (in the affirmative case) a rational parametrization of each component.

---

1. Compute the normal vector $\mathcal{N}(t)$ of $\mathcal{P}(t)$. IF $\|\mathcal{N}(t)\| \in \mathbb{K}(\bar{t})$ THEN RETURN $\mathcal{O}_d(\mathcal{C})$ has two rational components parametrized by $\mathcal{P}(t) \pm \frac{d}{\|\mathcal{N}(t)\|}\mathcal{N}(t)$.
2. Determine $\mathcal{G}_{\mathcal{P}}^{\mathcal{O}}(\mathcal{C})$, and decide whether $\mathcal{G}_{\mathcal{P}}^{\mathcal{O}}(\mathcal{C})$ is rational.
3. IF $\mathcal{G}_{\mathcal{P}}^{\mathcal{O}}(\mathcal{C})$ is not rational THEN RETURN no component of $\mathcal{O}_d(\mathcal{C})$ is rational.
4. Else compute a proper parametrization $\mathcal{R}(t) = (\tilde{R}(t), R(t))$ of $\mathcal{G}_{\mathcal{P}}^{\mathcal{O}}(\mathcal{C})$ and RETURN that $\mathcal{O}_d(\mathcal{C})$ is rational and that $\mathcal{Q}(t) = \mathcal{P}(\tilde{R}(t)) + \frac{2 \, d \, R(t)}{N_2(\tilde{R}(t))(R(t)^2+1)}\mathcal{N}(\tilde{R}(t))$ where $\mathcal{N} = (N_1, N_2)$, parametrizes $\mathcal{O}_d(\mathcal{C})$.

The Conchoid Rationality Problem

In [16], it is proved that conchoids having all their components rational can only be generated by rational curves. Moreover, it is shown that reducible conchoids to rational curves have always their two components rational (double rationality). From these results, one deduces that the rationality of the conchoid component, to a rational curve, does depend on the base curve and on the focus but not on the distance. To approach the problem we use similar ideas to those for offsets introducing the notion of *reparametrization curve* as well as the notion of *rdf parametrization*. The rdf concept allows us to detect the double rationality while the reparametrization curve is a much simpler curve than the conchoid, directly computed from the input rational curve and the focus, and that behaves equivalently as the conchoid in terms of rationality. As a consequence of these theoretical results [16] provides an algorithm to solve the problem. The algorithm analyzes the rationality of all the components of the conchoid and, in the affirmative case, parametrizes them. The problem of detecting the focuses from where the conchoid is rational or with two rational components is, in general, open.

We say that a rational parametrization $\mathcal{P}(t) = (P_1(t), P_2(t)) \in \mathbb{K}(t)^2$ of $\mathcal{C}$ is at *rational distance to the focus* $A = (a, b)$ if $(P_1(t) - a)^2 + (P_2(t) - b)^2 = m(t)^2$, with $m(t) \in \mathbb{K}(t)$. For short, we express this fact saying that $\mathcal{P}(t)$ is rdf or A-rdf if we need to specify the focus. On the other hand, we define the *reparametrization curve of the conchoid* $\mathfrak{C}_d^A(\mathcal{C})$ *associated to* $\mathcal{P}(t)$, denoted by $\mathcal{G}_{\mathcal{P}}^{\mathfrak{C}}(\mathcal{C})$, as the primitive part with respect to x_2 of the numerator of $-2x_2(P_1(x_1) - a) + (x_2^2 - 1)(P_2(x_1) - b)$.

Algorithm: conchoid parametrization

- GIVEN: a proper rational parametrization $\mathcal{P}(t)$ of a plane curve $\mathcal{C}$ in $\mathbb{K}^2$, a focus $A = (a, b)$, and $d \in \mathbb{C}$.
- DECIDE: whether the components of the conchoid $\mathfrak{C}_d^A(\mathcal{C})$ are rational.
- DETERMINE: (in the affirmative case) a rational parametrization of each component.

---

1. Compute $\mathcal{G}_{\mathcal{P}}^{\mathfrak{C}}(\mathcal{C})$.
2. If $\mathcal{G}_{\mathcal{P}}^{\mathfrak{C}}(\mathcal{C})$ is reducible RETURN that $\mathfrak{C}_d^A(\mathcal{C})$ is double rational and that $\mathcal{P}(t) + \frac{d}{\pm\|\mathcal{P}(t)-A\|}(\mathcal{P}(t) - A)$ parametrize the two components.
3. Check whether the genus of $\mathcal{G}_{\mathcal{P}}^{\mathfrak{C}}$ is zero. If not, RETURN that $\mathfrak{C}_d^A(\mathcal{C})$ is not rational.
4. Compute a proper parametrization $(\phi_1(t), \phi_2(t))$ of $\mathcal{G}_{\mathcal{P}}^{\mathfrak{C}}$ and RETURN that $\mathfrak{C}_d^A(\mathcal{C})$ is rational and that $\mathcal{P}(\phi_1(t)) + \frac{d}{\pm\|\mathcal{P}(\phi_1(t))-A\|}(\mathcal{P}(\phi_1(t)) - A)$ parametrizes $\mathfrak{C}_d^A(\mathcal{C})$.

We can note that the rationality of the both constructions is not equivalent. For instance, if $\mathcal{C}$ is the parabola of equation $y_2 = y_1^2$, that can be parametrized as (t, t^2), the offset at distance d is rational. However, the rationality of the conchoid of the parabola depends on the focus.

2   Implementation of Conchoid and Offset Maple Packages and Help Pages

In this section, we present the creation of two packages in the computer algebra system Maple, that we call **Conchoid** and **Offset**. These packages compute the implicit equation, and analyze the rationality and the reducibility of conchoids and offset curves respectively, providing rational parametrizations in case of genus zero. In addition, it allows us to display plots. These packages consist in several procedures that are based on the above parametrization algorithms.

In the following, we give a brief description of the procedures and we show one of the help pages for one of the Maple functions. The procedure codes and packages are available in

http://www.euitt.upm.es/uploaded/docs_personales/sendra_pons_juana/
offsets_conchoids/Offset.zip

2.1   Procedures of the Conchoid Package

getImplConch This procedure determines the implicit equation of the conchoid of an algebraic plane curve, given implicitely, at a fixed focus and a fixed distance. For this purpose, we use Gröbner basis to solve the system of equations consisting on the circle centered at generic point of the initial curve C and radius d, the straight line from the focus A to the generic point of the initial curve C, and the initial curve C.

getParamConch Firstly this procedure checks whether the conchoid of a rational curve is irreducible or it has two rational components. For this purpose, we analyze whether a proper rational parametrization of the initial curve is RDF. In affirmative case, the procedure outputs a message indicating reducibility (the conchoid has two rational components) and a rational parametrization for each component is displayed. Otherwise, the conchoid is irreducible and the reparametrization curve is computed in order to study its rationality. In the affirmative case, it provides a rational parametrization by means of a rational parametrization of the reparametrizing curve and it outputs a message indicating irreducibility and rationality.

plotImplConch This procedure computes the conchoids curve using *getImplConch* procedure, and then it plots both the initial curve and its conchoid within the coordinates axes interval $[-a, a] \times [-a, a]$.

2.2   Procedures of the Offset Package

ImplicitOFF This procedure determines the implicit equation of the offset of a rational algebraic plane curve, given parametrically, at a fixed distance. For this purpose, since the algebraic system has three variables and one parameter (namely the distance), instead of Gröbner basis we simplify the computation by using resultants to solve the system of equations consisting on the circle centered at a generic point of the initial curve C and radius d, and the normal line at each point of C.

OFFparametric This procedure analyzes the rationality of the offset of a rational plane curve. For this purpose, first it decides whether the offset is irreducible or it has two rational components. In case of reducibility, the procedure outputs a rational parametrization for each component, using the RPH concept. Otherwise, it checks whether the offset is rational or not. In the affirmative case, it provides a rational parametrization by means of a rational parametrization of the reparametrizing curve.

$\boxed{\textbf{OFFplot}}$ This procedure computes the offset curve at a generic distance, d, and then replaces d with fixed value, $dist$. Finally, it plots both the initial curve and its offset at a distance $dist$ within the coordinates axes interval $[-a, a] \times [-a, a]$.

Once we have implemented the Offset/Conchoid procedure in Maple, we have created two packages containing them, called *Conchoid* and *Offset*, respectively.

Theoretically, to compute the implicit equation of either the conchoid or the offset, we use the incidence varieties introduced in [1] and [15], respectively. In the definition of this incidence variety an equation, to exclude extraneous factors, is introduced such that the Zariski closure of the projection is exactly the offset/conchoid curve. Therefore, by the theorem of the closure, Gröbner basis computation, and resultant when possible, provides the correct equation. In addition, one has to take into account that we are dealing with generic conchoids and generic offsets and therefore the specialization of the Gröbner basis or the resultant may fail. Nevertheless, since we have only one parameter there are only finitely many specializations; In particular, $d = 0$ generates a bad specialization. Since $d = 0$ is not interesting from the geometric construction point of view we are excluding this case. In addition, we have created the help pages associated to the procedures.

3   Atlas of Conchoid and Offset Curves

In this section we illustrate the previous results applying the packages *Offset* and *Conchoid*. We analyze the rationality of the offset and the conchoid of several classical rational curves, and in the case of rationality we compute rational parametrizations. We give a table summarizing the main details of the process for each geometric construction, such as the degree of the implicit equation, rational character and rational parametrization in case of genus zero. In case of Conchoids, the rationality depends on the focus, therefore in the table we study the rationality for different focus position, distinguishing if the focus is on the base curve or not. We don't include the implicit equation of the reparametrizing curve because of space limitations. The implicit equations, plots and more details of the computation of these atlas are available by contacting with the corresponding author.

Table 1. Offsets Curves

Base Curve	Offset Degree	Rationality & Parametrization
Circle $x_2^2 + x_1^2 - 4$	4	Double Rational $\left(\pm \frac{(d \pm r)2t}{t^2+1}, \ \mp \frac{(d \pm r)(t^2-1)}{t^2+1} \right)$
Parabola $x_2 - x_1^2$	6	Rational $\left(\frac{(t^2-1)(-t^2-1+4\,dat)}{4at(t^2+1)}, \ \frac{t^6-t^4-t^2+1+32\,dt^3a}{16at^2(t^2+1)} \right)$
Hyperbola $\frac{x_1^2}{16} - \frac{x_2^2}{9} - 1$	8	Irreducible and non rational
Ellipse $\frac{x_1^2}{25} + \frac{x_2^2}{16} - 1$	8	Irreducible and non rational
Cardioid $(x_1^2 + 4x_2 + x_2^2)^2 - 16(x_1^2 + x_2^2)$	14	Rational $\left(\frac{(-9+t^2)(dt^6-117dt^4+3456t^3-1053dt^2+729d)}{(243t^2+27t^4+t^6+729)(t^2+9)}, \right.$ $\left. \frac{-18(dt^6-16t^5-21dt^4+864t^3-189dt^2-1296t+729d)t}{(243t^2+27t^4+t^6+729)(t^2+9)} \right)$
Three-leaved Rose $(x_1^2+x_2^2)^2 + x_1(3x_2^2 - x_1^2)$	14	Irreducible and non rational
Trisectrix of Maclaurin $x_1(x_1^2 + x_2^2) - (x_2^2 - 3x_1^2)$	10	Irreducible and non rational
Folium of Descartes $x_1^3 + x_2^3 - 3x_1x_2$	14	Irreducible and non rational
Tacnode $2x_1^4-3x_1^2x_2+x_2^2-2x_2^3+x_2^4$	20	Irreducible and non rational
Epitrochoid $x_2^4 + 2x_1^2x_2^2 - 34x_2^2 + x_1^4 - 34x_1^2 + 96x_1 - 63$	10	Irreducible and non rational
Ramphoid Cusp $x_1^4+x_1^2x_2^2-2x_1^2x_2-x_1x_2^2+x_2^2$	20	Irreducible and non rational
Lemniscata of Bernoulli $(x_1^2 + x_2^2)^2 - 4(x_1^2 - x_2^2)$	16	Irreducible and non rational

Table 2. Conchoids Curves

Curve $\mathcal{C}$	Focus Curve F / Focus Conch. A	Conchoid Parametrization
Circle	$A=F=(0,0)$ $A=(-2,0)\in\mathcal{C},$ $A=(-4,0)\notin\mathcal{C}$	$\left(\frac{-2(-1+t^2)\pm(1-t^2)}{1+t^2}, \frac{4t\pm2t}{1+t^2}\right)$ **DR** $\left(\frac{3t^4-12t^2+1}{1+2t^2+t^4}, \frac{2t(-3+5t^2)}{1+2t^2+t^4}\right)$ **R** **NR**
Parabola	$A=F=(0,1/4)$ $A\in\mathcal{C}, A=(0,0)$ $A=(0,-2)\notin\mathcal{C}$	$\left(t\pm\frac{4t}{1+4t^2}, t^2\pm\frac{4t^2-1}{1+4t^2}\right)$ **DR** $\left(\frac{2t+2t^3+1-2t^2+t^4}{(-1+t^2)(1+t^2)}, \frac{2t(2t+2t^3+1-2t^2+t^4)}{(-1+t)^2(1+t)^2(1+t^2)}\right)$ **R** **NR**
Hyperbola	$A=F=(5,0)$ $A=(-4,0)\in\mathcal{C}$ $A=(0,0)\notin\mathcal{C}$	$\left(\frac{-2(9+t^2)}{3t}\pm\frac{2(-t-6)(2t+3)}{45+24t+5t^2},\right.$ $\left.\frac{t^2-9}{2t}\pm\frac{3(t^2-9)}{45+24t+5t^2}\right)$ **DR** $\left(\frac{(45t^6+129t^4+311t^2+27)}{(1+t^2)(-9+t^2)(9t^2-1)}, \frac{2(-63+81t^4-82t^2)t}{(1+t^2)(-9+t^2)(9t^2-1)}\right)$ **R** **NR**
Ellipse	$A=F=(3,0)$ $A=(0,4)\in\mathcal{C}$ $A=(0,0)\notin\mathcal{C}$	$\left(\frac{5(t^2-1)}{t^2+1}\pm\frac{t^2-4}{t^2+1}, \frac{8t}{t^2+4}\pm\frac{4t}{t^2+4}\right)$ **DR** $\left(\frac{(1-t^2)(100t+100t^3+4t^4+17t^2+4)}{(4t^4+17t^2+4)(1+t^2)},\right.$ $\left.\frac{2(-58t^4+8t^6-58t^2+8-4t^5-17t^3-4t)}{(4t^4+17t^2+4)(1+t^2)}\right)$ **R** **NR**
Cardioid	$A=(0,0)\in\mathcal{C}$ $A=(-9,0)\notin\mathcal{C}$	$\left(\frac{-1024t^3}{(16t^2+1)^2}\pm\frac{-8t}{16t^2+1},\right.$ $\left.\frac{-128t^2(16t^2-1)}{(16t^2+1)^2}\pm\frac{1-16t^2}{16t^2+1}\right)$ **DR** **NR**
Three-leaved Rose	$A=(0,0)\in\mathcal{C}$ $A=(-2,0)\notin\mathcal{C}$	$\left(\frac{2(t^4-6t^2+9)t^2(t-1)(t+1)}{(t^4+2t^2+1)^2}, \frac{4t^3(t^4-6t^2+9)}{(t^4+2t^2+1)^2}\right)$ **R** **NR**
Trisectrix of Maclaurin	$A=(0,0)\in\mathcal{C}$ $A=(-4,0)\notin\mathcal{C}$	$\left(\frac{-2(-5t^2+2t^4+1)}{(t^4+2t^2+1)}, \frac{-4t(-5t^2+2t^4+1)}{(t^4+2t^2+1)(t^2-1)}\right)$ **R** **NR**
Folium of Descartes	$A=(0,0)\in\mathcal{C}$ $A=(-1,-1)\notin\mathcal{C}$	$\left(\frac{(-6t+6t^5+t^6-3t^4+3t^2-1+8t^3)(t-1)(t+1)}{(t^2+1)(t^6-3t^4+3t^2-1+8t^3)},\right.$ $\left.\frac{2(-6t+6t^5+t^6-3t^4+3t^2-1+8t^3)t}{(t^2+1)(t^6-3t^4+3t^2-1+8t^3)}\right)$ **R** **NR**
Tacnode	$A=(0,0)\in\mathcal{C}$ $A=(0,1)\in\mathcal{C}$	**NR** **NR**
Epitrochoid	$A=(3,0)\in\mathcal{C}$ $A=(0,0)\notin\mathcal{C}$	$\left(\frac{-7t^4+288t^2+256}{(t^2+16)^2}\pm\frac{16-t^2}{t^2+16},\right.$ $\left.\frac{-16t(5t^2-16)}{(t^2+16)^2}\pm\frac{(-8t)}{t^2+16}\right)$ **DR** **NR**
Ramphoid Cusp	$A=(0,0)\in\mathcal{C}$ $A=(-1,-1)\notin\mathcal{C}$	**NR** **NR**
Lemniscata of Bernoulli	$A=(-1,-1)\in\mathcal{C}$ $A=(-2,0)\in\mathcal{C}$	**NR** **NR**

DR Double Rational, **R** Rational, **NR** Irreducible and Non Rational

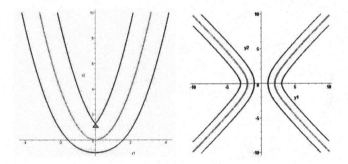

Fig. 2. Left: Parabola and the offset at $d = 2$. Right: Hyperbola and the offset at $d = 1.5$.

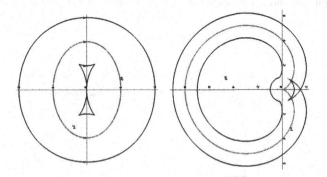

Fig. 3. Left: Ellipse and the offset at $d = 1$. Right: Cardioid and the offset at $d = 1$.

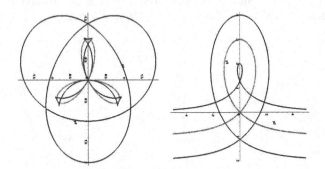

Fig. 4. Left: Three-leaved Rose and the offset at $d = 1$. Right: Trisectrix of Maclaurin and the offset at $d = 1$.

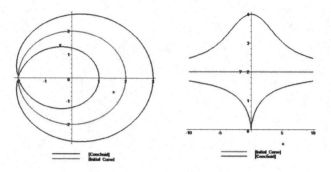

Fig. 5. Left: Circle and the conchoid at $A = (-2,0)$ and $d = 1$ (Limaçon of Pascal). Right: Straight line and the conchoid at $A = (0,0)$ and $d = 2$ (Conchoid of Nicomedes).

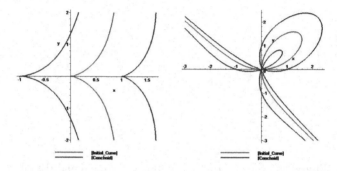

Fig. 6. Left: Conchoid of Sluze and the conchoid at $A = (-2,0)$ and $d = 1$. Right: Folium of Descartes and the conchoid at $A = (0,0)$ and $d = 2$.

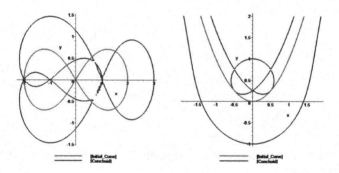

Fig. 7. Left: Lemniscata of Bernoulli and the conchoid at $A = (-2,0)$ and $d = 1$. Right: Parabola and the conchoid at $A = (0,1/4)$ and $d = 1$.

Acknowledgements. Corresponding Author supported by the Spanish Ministerio de Economia y Competitividad under the Project MTM2011-25816-C02-01.

References

1. Arrondo, E., Sendra, J., Sendra, J.R.: Parametric Generalized Offsets to Hypersurfaces. Journal of Symbolic Computation 23(2–3), 267–285 (1997)
2. Arrondo, E., Sendra, J., Sendra, J.R.: Genus Formula for Generalized Offset Curves. Journal of Pure and Applied Algebra 136(3), 199–209 (1999)
3. Farouki, R.T., Neff, C.A.: Analytic Properties of Plane Offset Curves. Comput. Aided Geom. Des. 7, 83–99 (1990)
4. Farouki, R.T., Neff, C.A.: Algebraic Properties of Plane Offset Curves. Comput. Aided Geom. Des. 7, 100–127 (1990)
5. Hoffmann, C.M.: Geometric and Solid Modeling. Morgan Kaufmann Publis. (1993)
6. Kerrick, A.H.: The limaçon of Pascal as a basis for computed and graphic methods of determining astronomic positions. J. of the Instit. of Navigation 6, 5 (1959)
7. Menschik, F.: The hip joint as a conchoid shape. J. of Biomechanics 30(9) (September 1997) 971-3 9302622
8. Peternell, M., Gruber, D.: Conchoid surfaces of quadrics. Journal of Symbolic Computation (2013), doi:10.1016/j.jsc, 07.003
9. Peternell, M., Gruber, D., Sendra, J.: Conchoid surfaces of rational ruled surfaces. Comp. Aided Geom. Design 28, 427–435 (2011)
10. Peternell, M., Gruber, D., Sendra, J.: Conchoid surfaces of spheres. Comp. Aided Geom. Design 30(1), 35–44 (2013)
11. Peternell, M., Gotthart, L., Sendra, J., Sendra, J.: The Relation Between Offset and Conchoid Constructions. arXiv:1302.1859v2 [math.AG] (June 10, 2013)
12. Peternell, M., Pottmann, H.: A Laguerre geometric approach to rational offsets. Computer Aided Geometric Design 15, 223–249 (1998)
13. Sendra, J., Sendra, J.R.: Algebraic Analysis of Offsets to Hypersurfaces. Mathematische Zeitschrif 234, 697–719 (2000)
14. Sendra, J., Sendra, J.R.: Rationality Analysis and Direct Parametrization of Generalized Offsets to Quadrics. Applicable algebra in Engineering, Communication and Computing 11, 111–139 (2000)
15. Sendra, J., Sendra, J.R.: An Algebraic Analysis of Conchoids to Algebraic Curves. Applicable Algebra in Engineering, Communication and Computing 19, 413–428 (2008)
16. Sendra, J., Sendra, J.R.: Rational parametrization of conchoids to algebraic curves, Applicable Algebra in Engineering. Communication and Computing 21(4), 413–428 (2010)
17. Sendra, J.R., Sevilla, D.: Radical Parametrizations of Algebraic Curves by Adjoint Curves. Journal of Symbolic Computation 46, 1030–1038 (2011)
18. Sendra, J.R., Sevilla, D.: First Steps Towards Radical Parametrization of Algebraic Surfaces. Computer Aided Geometric Design 30(4), 374–388 (2013)
19. Sultan, A.: The Limaçon of Pascal: Mechanical Generating Fluid Processing. J. of Mechanical Engineering Science 219(8), 813–822 (2005) ISSN.0954-4062
20. Weigan, L., Yuang, E., Luk, K.M.: Conchoid of Nicomedes and Limaçon of Pascal as Electrode of Static Field and a Wavwguide of High Frecuency Wave. Progres. In: Electromagnetics Research Symposium, PIER, vol. 30, pp. 273–284 (2001)

Algorithmic Aspects of Theory Blending

Maricarmen Martinez[1], Ulf Krumnack[2], Alan Smaill[3], Tarek Richard Besold[2],
Ahmed M.H. Abdel-Fattah[2,4], Martin Schmidt[2], Helmar Gust[2],
Kai-Uwe Kühnberger[2], Markus Guhe[3], and Alison Pease[5]

[1] Department of Mathematics, Universidad de los Andes, Bogotá, Colombia
[2] Institute of Cognitive Science, University of Osnabrück, Osnabrück , Germany
[3] School of Informatics, University of Edinburgh, Edinburgh, Scotland, UK
[4] Computer Science Division, Faculty of Science, Ain Shams University, Cairo, Egypt
[5] School of Computing, University of Dundee, Dundee, Scotland, UK

Abstract. In Cognitive Science, conceptual blending has been proposed as an important cognitive mechanism that facilitates the creation of new concepts and ideas by constrained combination of available knowledge. It thereby provides a possible theoretical foundation for modeling high-level cognitive faculties such as the ability to understand, learn, and create new concepts and theories. This paper describes a logic-based framework which allows a formal treatment of theory blending, discusses algorithmic aspects of blending within the framework, and provides an illustrating worked out example from mathematics.

1 Introduction

Since its introduction, the theoretical framework of Conceptual Blending (CB) has gained popularity as alleged submechanism of several complex high-level cognitive capacities, such as counterfactual reasoning, analogy, and metaphor [2]. While there is a growing body of work trying to conceptually relate CB to several facilities at the core of cognition, there currently are very few (if any) fully worked out formal or algorithmic accounts. Still, if only some of the assumptions made about the importance of blending mechanisms within human cognition and intelligence turn out to be reliable, a complete and implementable formalization of CB and its defining characteristics would promise to trigger significant development in artificial intelligence.

An early formal account on CB, especially influential to our approach, is the classical work by Goguen using notions from algebraic specification and category theory [3].

This version of CB is depicted in Figure 1, where a blend of two inputs I_1 and I_2 is shown. Each node in the figure stands for a representation of a concept or conceptual domain as a theory (set of axioms) in a formal language. We'll call the nodes "spaces", so to avoid terms with strong semantical load such as "concept" or "conceptual domain". Each arrow in the figure stands for a morphism, that is, a change-of-language *partial* function that translates at least part of the axioms from its domain into axioms in its codomain, preserving their structure. Now, while in practice all formal languages of interest have a established semantics and the morphisms are therefore intended to act as partial interpretations of one theory into another, Goguen's presentation of CB stays at the syntactic level, which more directly lends itself to computational treatment.

G.A. Aranda-Corral et al. (Eds.): AISC 2014, LNAI 8884, pp. 180–192, 2014.

The same will apply to our own approach. Given input spaces I_1 and I_2 and a generalization space G that encodes some (ideally all) of the structural commonalities of I_1 and I_2, a blend diagram is completed by a blend space B and morphisms from I_1 and I_2 to B such that the diagram (weakly!) commutes. This means that if two parts of I_1 and I_2 are translated into B and in addition are identified as 'common' by G, then they must be translated into exactly the same part of B (whence the term 'blend').

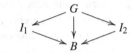

Fig. 1. Goguen's version of concept blending (cf. [3])

A standard example of CB, discussed in [3] and linked to earlier work on computational aspects of blending in cognitive linguistics (see, e.g., [11]), is that of the possible blends of HOUSE and BOAT into both BOATHOUSE and HOUSEBOAT (as well as other less-obvious blends). Parts of the spaces of HOUSE and BOAT can be structurally aligned (e.g. a RESIDENT LIVES-IN a HOUSE; a PASSENGER RIDES-ON a BOAT). Conceptual blends are created by combining features from the two spaces, while respecting the constructed alignments between them. Newly created blend spaces are supposed to coexist with the original spaces: we still want to maintain the spaces of HOUSE and BOAT.

A still unsolved question is to find criteria to establish whether a blend is better than other candidate blends. This question has lead to the formulation of various competing optimality principles in cognitive linguistics (cf. [2]). While several of them involve semantic aspects that escape Goguen's and our own treatment of CB, other principles can be reasonably approached even from a more syntactic framework. For example, there is the *Web Principle* (maintain as tight connections as possible between the inputs and the blend), the *Unpacking Principle* (one should be able to reconstruct the inputs as much as possible, given the blend), and the Topology Principle (the components of the blend should have similar relations to those that their counterparts hold in the input spaces). These three principles, taken as a package, can be interpreted in terms of Figure 1 as demanding that the morphisms should preserve as much representational structure as possible. For example, one can notice that Figure 1 looks like the diagram of a pushout in category theory. Goguen actually argued against forcing the diagram of every blend to be a pushout [3], but he did claim that some forms of a pushout construction (in a $\frac{3}{2}$-category) capture a notion of *structural optimality* for blends.

We will propose two alternative competing criteria for structural blend optimality that also work in the spirit of the Web, Unpacking, and Topology principles, and an algorithmic method for performing blending guided by those principles. We will use HDTP, a framework for computational analogy making between *first-order* theories, in order to obtain the generalization spaces G. Accordingly, our presentation here will be restricted to CB over first-order theories. The paper is structured as follows: we first introduce the formal framework we use to model blending processes, and then propose our algorithmic description of blending. As proof of concept, along the paper we present a worked out example from mathematics. The paper finishes with some concluding remarks, a review of related work, and an outlook for future research.

2   Our Framework

Our approach is based on the *Heuristic-Driven Theory Projection* (HDTP, [10]), a framework for computing analogical relations between two input spaces presented as axiomatizations in some many-sorted first-order languages. HDTP proceeds in two phases: in the *mapping phase*, the source and target spaces are compared to find structural commonalities and a generalized space G is created, which subsumes the matching parts of both spaces. In the *transfer phase*, unmatched knowledge in the source space can be mapped to the target space to establish new hypotheses (Figure 2). For our current purposes we will only need the mapping mechanism and replace the transfer phase by a new blending algorithm, so instead of talking about source and target spaces, from now on we will refer to the input spaces as the 'left' and 'right' spaces (L and R). This convention is meant to be merely a mnemonic relating to our diagrams and not an indication that one space has priority over the other (since we don't need transfer anymore).

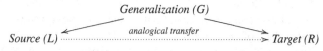

Source (L)   analogical transfer   Target (R)

Generalization (G)

Fig. 2. HDTP's overall approach to creating analogies (cf. [10])

During the mapping phase in HDTP, pairs of formulae from L and R are *anti-unified*, resulting in a generalization theory G that reflects common aspects of the input spaces. Anti-unification [9] is a mechanism that finds least-general anti-unifiers of expressions (formulae or terms). An anti-unifier of A and B is an expression E such that A and B can be obtained from E via substitutions. E is a least-general anti-unifier of A and B if it is an anti-unifier and the only susbtitutions on E that yield anti-unifiers of A and B act as trivial renamings of the variables in E. As it happens, first-order anti-unification (where only first-order substitutions are allowed) is not powerful enough to produce the generalizations needed in HDTP, so a special form of *higher-order* anti-unification is used where, under certain conditions, symbols of relation and function can also be included in the domain of substitutions (see [10] for the details). The generalized theory G can be projected into the original spaces by higher-order substitutions which are computed by HDTP during anti-unification. We will say that a formula is *covered* by G if it is in the image of this projection; otherwise it is *uncovered*.

Example 1. We will use a working example in this paper based on the theories L and R from Table 1, which describe basic properties of the standard order and addition of the natural numbers (starting from 1) and the non-negative rationals, respectively. All the axioms are implicitly universally quantified, and $x <_i y$ abbreviates $\neg(y \leq_i x)$. The table also shows a generalization theory G over the signature is $\{a, \leq, +\}$. G reflects the fact that axiom (Li) is structurally like (Ri) when $1 \leq i \leq 6$. Upon applying the left and right substitutions to G, we'll get the first six L-axioms and the first six R-axioms, respectively, which are the covered formulas in this example.

In HDTP, any two formulae (or terms) from the input spaces that are generalized (i.e. anti-unified) to the same expression in G are considered to be analogical. In analogy making, the analogical relations are used in the transfer phase to translate uncovered

Table 1. The two axiomatizations and the first generalization G used in the worked example. G comes together with a **left substitution** $\{a \mapsto 1, \leq \mapsto \leq_L, + \mapsto +_L\}$ and a **right substitution** $\{a \mapsto 0, \leq \mapsto \leq_R, + \mapsto +_R\}$ from which L and R can be recovered.

Axiomatization L		Axiomatization R		Generalization G	
$x \leq_L x$	(L1)	$x \leq_R x$	(R1)	$x \leq x$	(G1)
$x \leq_L y \wedge y \leq_L z \to x \leq_L z$	(L2)	$x \leq_R y \wedge y \leq_R z \to x \leq_R z$	(R2)	$x \leq y \wedge y \leq z \to x \leq z$	(G2)
$x \leq_L y \vee y \leq_L x$	(L3)	$x \leq_R y \vee y \leq_R x$	(R3)	$x \leq y \vee y \leq x$	(G3)
$1 \leq_L x$	(L4)	$0 \leq_R x$	(R4)	$a \leq x$	(G4)
$x +_L y = y +_L x$	(L5)	$x +_R y = y +_R x$	(R5)	$x + y = y + x$	(G5)
$(x +_L y) +_L z = x +_L (y +_L z)$	(L6)	$(x +_R y) +_R z = x +_R (y +_R z)$	(R6)	$(x + y) + z = x + (y + z)$	(G6)
$\neg (x +_L 1 \leq_L x)$	(L7)	$x +_R 0 = x$	(R7)		
$x \leq_L y \wedge y \leq_L x +_L 1 \to y = x \vee y = x +_L 1$	(L8)	$x <_R y \to \exists z : x <_R z \wedge z <_R y$	(R8)		

facts from the source to the target space, while blending combines uncovered facts from both spaces. Thus, the blending process can build on the generalization and substitutions provided by the analogy engine, and analogy can be considered a special case of blending.

There are two extreme cases of CB, depending on the portion of the input theories covered by G. The first case (left side of Figure 3) occurs when the input spaces are isomorphic, meaning that there is a bijective morphism that simply renames the signature symbols of the language of L onto the symbols of R. In that case, all formulae of the theories can be generalized and are completely covered by G, and the resulting blend will be isomorphic to both of them [1]. The other extreme (right side of Figure 3) occurs

Fig. 3. The two extreme cases of input spaces, along with their generalizations and blends

when no formulae can be aligned and therefore the generalized theory G is empty, so no formulae of the input theories are covered. In this case, a blend can always be obtained by taking the (possibly inconsistent) disjoint union of the input theories. In practice, neither of the two extreme cases is of real interest. The interesting *proper blends* arise when only parts of the input theories are covered by G. In fact, one can adjust the blend by changing the generalization, either by removing formulae from G and so reducing its coverage, or by choosing altogether another G which associates different formulae.

Given G, the theories L and R can be split into their (non-empty) covered parts L^+ and R^+ and uncovered parts L^- and R^-. The covered parts are fully analogical, i.e. basically isomorphic, and make up the core of the a blend B based on G. The uncovered parts reflect the idiosyncratic aspects of the spaces, which we would ideally want to integrate into B. However, due to the identifications induced by G, adding all this to B may result in an inconsistent theory. To preserve consistency, we may be forced to

---

[1] HDTP is syntax-based, but has some "re-representation" abilities by which formulae derived from the axioms may be used in the mapping phase if the original axiomatizations don't yield a good analogical relation (cf. [10, pp. 258]). Thus, in some cases, two formally different but semantically equivalent axiomatizations may not result in an empty generalization.

consider only consistent subsets of this ideal, fully inclusive, blend. In view of this, we propose the following two optimality principles: IP renders a version of the Web and Topology principles formulated in the introduction, while CP supports the Unpacking Principle.

> COMPRESSION PRINCIPLE (CP): aim for blend diagrams in which B is as compressed as possible, that is, where as many signature symbols aligned by G as possible are actually integrated as a single symbol in B.

> INFORMATIVENESS PRINCIPLE (IP): aim for blend diagrams in which B is as informative as possible, i.e., it includes a maximally consistent subset of the potentially merged formulae (obtained by taking the union of the input theories and then collapsing pairs of signature symbols that have been identified by the analogy into one unified symbol).

3   Theory Blending Algorithm

Now we tackle the problem of algorithmically finding a list of *optimal* blends, given two input theories L and R over first-order signatures Σ_L, Σ_R, respectively. A blend is *optimal* if it is consistent and as maximally compressed and informative as possible. An unconstrained way to do this leads to an explosion of possibilities to be tried, so good heuristics are needed in order to choose which possibilities to test first. We propose to proceed according to the following general steps:

1. Generalization: Using the HDTP mapping phase, compute a generalization G that is as strong as possible (i.e identifies as many symbols as possible) together with its associated substitutions[2]. As an example, see Table 1 and Example 1.

2. Identification: Build the blend signature Σ_B by taking the 'union' of Σ_L and Σ_R and collapsing each pair of symbols aligned by G to only one of them. Regardless of how the collapsing is done, at the end the algorithm will produce the same blends, modulo partial renamings of identified symbols[3]. In what follows, we will simply choose the symbol from Σ_R when collapsing a pair. Thus, for the case of Table 1, Σ_B will coincide with Σ_R, since no symbol in Σ_L is uncovered by the left substitution.

3. Blending: Construct the set of all formulae over Σ_B that might be part of a blend. This will consist of every formula in R^+, the covered part of R, plus every formula in the uncovered parts of R and L, $Ax = Tr(L^-) \cup R^-$. Here Tr is the (partial) translation function that maps symbols from Σ_L to corresponding symbols from Σ_R according to the generalization G, so ensuring that all formulas of Ax are build over signature Σ_B. The set Ax corresponding to the example in Table 1 is listed in the leftmost column of Table 2, which also shows all the candidate blends for this particular generalization G.

Back to the general setting, the set $R^+ \cup Ax$ would be the ideal blend, but it might be inconsistent. So in this step we also compute the set MaxCon of maximal consistent blends B such that $R^+ \subseteq B \subseteq R^+ \cup Ax$. For the running example, this involves exploring the lattice of theories depicted in Figure 4.

---

[2] A simplified version of HDTP is used, where substitutions must preserve the arity of symbols.

[3] The algorithm might in principle be extended by producing for each discovered optimal blend all of its "mirror" blends, obtained by renamings.

The user of the algorithm decides now if the produced blends are good enough or the search must continue. In the first case we stop. If not, go to the next step which will need the set MinInc of minimally inconsistent subsets of $R^+ \cup Ax$ that extend R^+.

4. Relaxation: Reduce the set of symbols covered by the generalization by shrinking G (some simple heuristics for this step are given below). Return to step 2.

Now we discuss how steps 3 and 4 can be implemented (steps 1 and 2 are obtained from HDTP). We use a simple procedure COMPUTEBLENDS which, besides the sets R^+ and Ax introduced above, needs a list Init of initial blend candidates (so each element of Init extends R^+). Init must have the property that every possible blend based on the current generalization is either a superset or a subset of one of the elements of Init. This, plus the way in which Init will be changed in the relaxation phase (more on this below) guarantees that the algorithm will find *all* the optimal blends if never asked to stop the search (at the end of step 3). At the very beginning of the process (step 1 above) Init can be initialized, for example, to be the set of theories that extend R^+ (a different choice will be used later in our worked example). When a relaxation is needed (step 4 above) a new set Init is computed from MaxCon and MinInc (more on this later). There is a fourth parameter ('direction') which is used to direct the search as explained soon.

proc COMPUTEBLENDS(R^+, Ax, Init, direction)
    global MaxCon := $\emptyset$; **global** MinInc := $\emptyset$
    foreach $T \in$ Init **do** EXPLORE(R^+, Ax, T, direction) **end foreach**
end proc

The first thing to do is to initialize as empty two global sets MaxCon and MinInc that will keep at all times during the search the largest consistent theories and the smallest inconsistent theories that have been found up to the moment. After this initialization, the procedure enters into a loop in which for each initial theory T in Init, the procedure EXPLORE will populate MaxCon and MinInc. After execution, all blends that contain T or are contained in T, will be *"classified correctly"* by MaxCon and MinInc, i.e. they will be subsumed by some theory in MaxCon if they are consistent, and they will subsume some theory from MinInc if they are inconsistent (cf. Lemma 1 below). When the loop ends, MaxCon determines precisely the optimal blends.

proc EXPLORE(R^+, Ax, T, direction)

    if $T \notin \downarrow$MaxCon $\cup \uparrow$MinInc **then**
        if T is consistent **then** MaxCon := $\{T\} \cup \{M \in$ MaxCon $\mid M \not\subseteq T\}$
        else MinInc := $\{T\} \cup \{M \in$ MinInc $\mid T \not\subseteq M\}$ **endif**
    endif
    if $T \in \downarrow$MaxCon **and** (direction $\in \{up, both\}$) **then**
        foreach $Axiom \in (Ax \setminus T)$ **do** EXPLORE(R^+, Ax, $T \cup \{Axiom\}$, **up**) **end foreach**
    else if $T \in \uparrow$MinInc **and** (direction $\in \{down, both\}$) **then**
        foreach $Axiom \in T \setminus R^+$ **do** EXPLORE(R^+, Ax, $T \setminus \{Axiom\}$, **down**) **end foreach**
    endif
end proc

Here, $\uparrow C$ denotes the set of theories that contain some theory from C and $\downarrow C$ denotes the set of theories that are contained in some theory from C; $\updownarrow C$ is $\uparrow C \cup \downarrow C$. As first step in EXPLORE, if T is not yet classified by MaxCon or MinInc, consistency of T is checked and MaxCon or MinInc are updated accordingly. In any case, if T is consistent (inconsistent), a recursive upwards (downwards) search towards extensions (subsets) of T is initiated. These upward and downward searches are performed unless the direction parameter prohibits them. The calls to EXPLORE made when working with the first, strongest generalization use always the direction *both*, with the effect that upwards and downwards searches are allowed. In the case of calls to EXPLORE after a 'relaxation' has been made, the direction is set to *up* (the reasons for this will be explained later)[4].

The above claims about EXPLORE follow from the next result, in which R^+ and Ax are fixed and the words "theory blend" refer to sets T such that $R^+ \subseteq T \subseteq R^+ \cup Ax$. Also, we will say that *MaxCon and MinInc classify correctly* if all the elements of MaxCon are consistent theory blends and all elements of MinInc are inconsistent theory blends.

Proposition 1. *The following pre- and post conditions hold true of the operation of* EXPLORE *(R^+, Ax, T, direction), for all theory blends T:*
(1) If all consistency checks can be accomplished, the procedure will terminate.
(2) If MaxCon and MinInc classify correctly before executing EXPLORE, *then the same holds afterwards.*
(3) If a theory blend B is classified correctly by MaxCon and MinInc before executing EXPLORE, *then the same holds after calling* EXPLORE.
(4) If direction $= up$ and MaxCon and MinInc classify correctly before executing EX-PLORE, *then $\uparrow T$ is classified correctly by MaxCon and MinInc after calling* EXPLORE.
(5) If direction $= up$ and MaxCon and MinInc classify correctly before executing EXPLORE, *then $\downarrow T$ is classified correctly by MaxCon and MinInc after calling* EXPLORE.
(6) If direction $= both$ and MaxCon and MinInc classify correctly before executing EX-PLORE, *then $\updownarrow T$ is classified correctly by MaxCon and MinInc after calling* EXPLORE.

Proof. To show (1) notice first that the recursion will only occur with strictly larger (direction = up) or strictly smaller (direction = down) values for T. As the size of T is limited by R^+ and $R^+ \cup Ax$ the claim follows. (2) follows directly, as MaxCon is only changed when a consistent blend T is added. The case for MinInc is analogous. (3) Let B be a consistent blend. By assumption $B \in \downarrow MaxCon$ before executing EXPLORE. MaxCon is only changed if T is consistent but $T \notin MaxCon$, in which case it will become $\{T\} \cup \{M \in MaxCon | M \not\subseteq T\}$. Now either $B \subseteq T$ or $B \subseteq M \in MaxCon$ with $M \not\subseteq T$. In both cases B is classified correctly by the new MaxCon. (4) We proceed by induction on the cardinality of $Ax \setminus T$. If T is inconsistent, no recursive call to EXPLORE is made. If $T \in \uparrow MinInc$ there is nothing to prove. If $T \notin \uparrow MinInc$, observe that T will be added to MinInc, so at the end of the procedure $\uparrow T$ will be classified correctly by *MaxCon* and *MinInc*. Now, if T is consistent and $T \notin \downarrow MaxCon$, then T will be added to MaxCon. Then, for each element A of $Ax \setminus T$, a call EXPLORE$(R^+, Ax, T \cup \{A\}, \textbf{up})$ will be made. By inductive hypothesis, after all these calls, every $\uparrow(T \cup \{A\})$ is classified correctly by *MaxCon* and *MinInc*, and so (since T is also classified correctly) $\uparrow T$ is classified correctly. (5) The argument is analogous to that for (4), now using induction on the cardinality of $T \setminus R^+$. (6) If T is consistent, an argument very close to that of (4)

---

[4] There are standard ways to improve the efficiency of the above procedure (using ordered lists, for example), but such discussion would lead us away from the main focus of this paper.

shows that $\uparrow T$ is classified correctly, so $T \subseteq T'$ for some $T' \in MaxCon$. Then $\downarrow T$ is classified correctly as well. A similar argument applies if T is inconsistent.   □

As our framework stands, the evaluation of blends in Step 3 and the decision to stop or continue with a relaxation, is a mandatory interactive step where the user decides. As for the relaxation step, if needed, it is important to find a good weakening of G a good set Init with which to continue to step 2. In principle, the framework allows for an interactive implementation where the user decides which weakened generalization to use next, or for an implementation that uses automated heuristics, such as building a weakened generalizations for which: (1) only one old symbol mapping is dropped, and (2) the fewest number of axioms become uncovered under the new generalization.

In any case, once a weakened generalization $\hat{G}$ has been fixed, the previously found MaxCon and MinInc sets are used to compute an appropriate new Init set, as follows. Let Tr and $\hat{T}r$ be the old and new translation functions. To form the set Init, for each T in MinInc (and optionally for every minimal extension of MaxCon) add to Init the theory that results from replacing in T every formula of the form $Tr(\phi)$ in R^- by $\hat{T}r(\phi)$. This new Init is good in that every optimal blend for the weakened generalization will be an extension of one the Init elements. This is why the exploration, after some relaxation has been made, can be constrained to be upwards only.

Our algorithm involves testing theories in first-order logic with equality for inconsistency; this is well-known to be undecidable in general. In our examples the inconsistencies will be discovered quickly[5], but in more elaborate situations, a resource-bounded check for inconsistency may model reasonably well the experience of mathematicians who can work productively with theories that are believed to be consistent and later revise their results in case an inconsistency is found. Research on Nelson Oppen methods (see [7] for a survey) reveals conditions under which the satisfiability and decidability of two theories is preserved when taking their union. The basic case requires the signatures of the two theories to be disjoint, but this can sometimes be relaxed. Some of these technical results might end up being useful to our work.

4   Worked Example

To illustrate the algorithm and suggest at least one improvement to it, we come back to take the theories shown in Table 1. Remember that L is based on the additive natural numbers (starting from 1) and L on the non-negative rational numbers. Thus, the notion of 'number' in L is discrete with least element 1, whereas in R it is dense with least element 0 (as the neutral element for addition). We will find all the optimal blends of L and R. The example shows that our approach isolates just a few optimal blends among many candidates, and that the short list includes (although not exclusively) the ones that one would expect a mathematician to judge as most interesting.

The first stage of the procedure was already partially described in the previous section. It explores the potential blends based on the generalization G of Table 1. Figure 4 shows a lattice of the blends and Table 1 lists the axioms of each candidate blend. Our set of initial theories will be formed by the minimal extensions of theory R and the minimal extensions of (the transferred version of) theory L. That is, Init$:= \{T1, T3, T7, T4\}$.

---

[5] HDTP and an beta implementation of the blending phase module are available on request. The blending module uses *prover9* to check for consistency.

Table 2. Formulae $L7t$ and $L8t$ result from transferring the uncovered formulae of axiomatization L, according to generalization G. The table shows some of the theories in the search space of possible blends. Maximal consistent theories are starred.

		TR	T1	T2	T3	T4	T5	T6	T7	T8	T9	TL
$x \leq_R x$	(R1)	X	X	X	X	X	X	X	X	X	X	X
$x \leq_R y \wedge y \leq_R z \to x \leq_R z$	(R2)	X	X	X	X	X	X	X	X	X	X	X
$x \leq_R y \vee y \leq_R x$	(R3)	X	X	X	X	X	X	X	X	X	X	X
$0 \leq_R x$	(R4)	X	X	X	X	X	X	X	X	X	X	X
$x +_R y = y +_R x$	(R5)	X	X	X	X	X	X	X	X	X	X	X
$(x +_R y) +_R z = x +_R (y +_R z)$	(R6)	X	X	X	X	X	X	X	X	X	X	X
$x +_R 0 = x$	(R7)	X	X			X	X	X	X		X	
$x <_R y \to \exists z : (x <_R z \wedge z <_R y)$	(R8)	X	X	X	X	X		X		X		
$\neg(x +_R 0 \leq_R x)$	(L7t)		X	X	X		X	X	X			X
$x \leq_R y \wedge y \leq_R x + 0 \to y = x \vee y = x +_R 0$	(L8t)				X	X		X	X	X	X	X
Consistent:		Y	N	Y*	N	Y*	N	N	N	Y	Y	Y*

The sets MaxCon and MinInc are initialized as empty and we start to explore the initial theories. The first is $T1$, which is inconsistent:

$$x +_R 0 = x \tag{R7}$$
$$\neg(x +_R 0 \leq_R x) \tag{L7t}$$
$$\neg(x \leq_R x) \tag{Substitution}$$
$$x \leq_R x \tag{R1}$$

The two last lines are clearly contradictory. The algorithm orders to add $T1$ to MinInc. However, knowing that the inconsistency arises from only the axioms $R1, R7$, and $L7t$, it is better to add the smaller $T5$ to MinInc than adding $T1$ itself. Thus, MinInc:= $\{T5\}$.

Now, as the algorithm prescribes, we recursively explore (downwards) every theory obtained from $T1$ by deleting one axiom. These theories are $TR, T2$, and $T5$: TR is consistent and $T5 \not\subseteq TR$, so MaxCon := $\{TR\}$; $T2$ is consistent, not contained in TR, and does not extend $T5$, then we update MaxCon := $\{TR, T2\}$; and $T5$ extends the only member of MinInc, so we do nothing. This ends the analysis of $T1$.

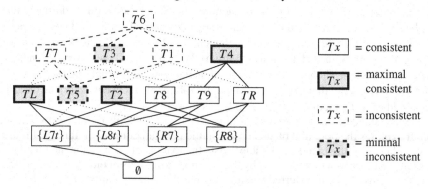

Fig. 4. A lattice of the 'blends' that appear in the given example

The second initial theory is $T3$. This theory is not a subset of TR or $T2$, and does not extend $T5$. In addition it is inconsistent, as shown by the third and last lines of the following proof, which uses all the axioms of $T3$ not covered by the generalization.

$$\neg(x +_R 0 \leq x) \tag{L7t}$$
$$\neg(x +_R 0 \leq x) \to \exists z : (x <_R z \wedge z <_R x +_R 0) \tag{R8}$$
$$x <_R z \wedge z <_R x +_R 0 \tag{FOL}$$
$$\neg(z \leq_R x) \wedge \neg(x + 0 \leq_R z) \tag{Def. $\leq_R$}$$
$$x \leq_R z \wedge z \leq_R x +_R 0 \tag{FOL + R3}$$
$$z = x \vee z = x +_R 0 \tag{MP with L8t}$$
$$z \leq_R x \vee x +_R 0 \leq_R x \tag{FOL + R1 + Def. $\leq_R$}$$

We update MinInc:= $\{T5, T3\}$, and recursively explore (downwards!) every theory obtained from $T3$ by erasing one axiom, namely $TL, T2$, and $T8$:

1. TL is consistent and does not extend TR nor $T2$, then MaxCon := $\{TR, T2, TL\}$. We are in the "downwards" mode, so we stop.
2. $T2$ is a member of MaxCon, so we stop.
3. $T8$ is consistent and not contained in a member of MaxCon. We set MaxCon := $\{TR, T2, TL, T8\}$. Again, we are in the "downwards" mode, so this branch stops.

This ends the analysis of $T3$, the second initial theory.

The third initial theory is $T7$, but the analysis of it stops immediately as it extends $T5 \in$ MinInc. We are left with the initial theory $T4$, which is consistent and not contained in Maxcon. Then Maxcon is updated by deleting the subsets of $T4$ (TR and $T8$) and adding $T4$: MaxCon := $\{T4, T2, TL\}$. Then we recursively explore (upwards) for possible consistent extensions of $T4$. The only proper extension of $T4$ is $T6$, which extends elements of MinInc. The first stage of the algorithm ends thus::

Solutions: $T2, T4$, and TL.   **Minimally inconsistent theories**: $T5$ and $T3$.

Note that TL is just a signature renaming of theory L, $T4$ a case of analogical transfer but not a *proper* blend, and $T2$ a proper blend intuitively describing the rationals larger than some nonzero number, which is not more interesting than the rationals starting with zero, to which L corresponds. It is then fair to assume that the user will decide to continue the search. In the second search stage, some of the contradictions found in stage 1 will be avoided by weakening the signature of the generalization in the relaxation step. The weakening heuristics described in the previous section suggest dropping the identification between 0 and 1, as this is the dropping that would diminish coverage the least. The new generalized theory changes only in that ($G4$) is not an axiom of it anymore. The result of transferring all of the axioms of axiomatization L to the R side involves the introduction of a new symbol of constant (1) to the R-side; cf. Table 3.

The set of initial theories will consist of the smallest versions, under the new signature, of the theories associated with the elements of MinInc from stage 1. More in detail, under the new signature there are four versions of each old theory Tj from the first stage. We call them $Tj0$, $Tj1$, $Tj2$, or $Tj3$ depending on which subset of $\{R4, L4tt\}$ they contain: $Tj0$ includes no element from $\{R4, L4tt\}$, $Rj1$ includes only $L4tt$, $Rj2$ includes only $R4$, and $Rj3$ includes the two axioms. Only some of these theories are shown in Table 3. Our set of initial theories in this stage will then be Init:= $\{T30, T50\}$. The sets MaxCon and MinInc are reset to the empty set.

Table 3. Formulae *Lxxx* result from transferring the uncovered formulae of *L* according to the weakened generalization that does not identify 0 and 1. Maximal consistent theories are starred.

		T30	T50	T51	T52	T53	T10	T11	T12	T13	T62	T72
(R1) − (R3), (R5), (R6)		X	X	X	X	X	X	X	X	X	X	X
$0 \leq_R x$	(R4)				X	X			X	X	X	X
$x +_R 0 = x$	(R7)		X	X	X	X	X	X	X	X	X	X
$x <_R y \to \exists z : (x <_R z \land z <_R y)$	(R8)	X						X	X	X	X	X
$1 \leq_R x$	(L4tt)				X		X		X		X	
$\neg(x +_R 1 \leq_R x)$	(L7tt)	X	X	X	X	X	X	X	X	X	X	X
$x \leq_R y \land y \leq_R x + 1 \to y = x \lor y = x +_R 1$	(L8tt)	X								X	X	X
Consistent:		N	Y	N	Y	N	Y	N	Y*	N	N	Y*

Every maximally compressed solution blend with respect to the new generalization must extend one of the initial theories. We explore each one of these initial theories in the "upwards" mode. We start with $T30$. This theory is inconsistent because the proof used in stage 1 to see that $T3$ is inconsistent still goes through when using 1 instead of 0 throughout, and $L7tt$ instead of $L7t$. We update MinInc := $\{T30\}$.

Then we test the second and last initial theory, $T50$. The theory is consistent but may not be maximal. We update MaxCon := $\{T50\}$, and explore $T50$'s minimal extensions:

1. $T51$ is inconsistent and does not extend $T30$, therefore MinInc := $\{T30, T51\}$.
2. $T10$ is consistent and extends $T50$. Set MaxCon := $\{T10\}$ and explore the three minimal extensions of $T10$, thus: $T60$ and $T11$ extend the elements $T30$ and $T51$ of MinInc, so nothing is done in these cases; and $T12$ is consistent and properly extends $T10$. Thus, we update MaxCon := $\{T12\}$ and test the minimal extensions of $T12$. There are only two cases of such a minimal extension: Adding $L4tt$ to $T12$ yields a theory that extends the element $T51$ of MinInc; and Adding $L8tt$ yields the theory $T62$, which is inconsistent because it extends $T30 \in$ MinInc.
3. $T70 = T50 \cup \{L8tt\}$ is consistent. So we update MaxCon := $\{T12, T70\}$, and explore the minimal extensions of $T70$. They are: $T60$ (which extends $T30 \in MinInc$), $T71$ (which extends $T51 \in MinInc$), and $T72$ (maximal consistent). After these explorations, MaxCon := $\{T12, T72\}$, and MinInc := $\{T30, T51\}$.
4. $T52$ is a subset of $T12 \in$ MaxCon, so we stop.

The second stage ends with new solutions $T12$ and $T72$, which, we claim, are the two mathematically interesting blends of the given theories: there are distinguished numbers 0 and 1, with 0 the unit for addition, and 1 strictly greater than 0; $T72$ is discrete, with a zero element immediately below 1, while $T12$ is dense, with a distinguished unit size.

5   Concluding Discussion

We presented a new algorithmic way of performing theory blending, based on the HDTP framework. Our approach is inspired by Goguen's treatment of CB, but differs from his in various aspects. First, our system generally outputs fewer blends focusing on maximal informativeness and compression as optimality criteria. By this we capture some aspects from [2]'s "optimality principles" for blends. Second, our algorithm uses only the weakenings of a fixed generalization, while Goguen seems to require the exploration of many (possibly mutually incompatible) starting generalizations. Our account also differs from that of [8], as there mappings "do not have to rely on similarity: they can present conflicts that are striking, surprising or even incongruous" [8, p. 90].

Our approach performs CB as theory blending. It therefore is especially appealing for applications in mathematics (such as the automated creation of mathematical concepts and conjectures) and logic-based AI. We demonstrated how traditional optimality criteria for CB can be spelled out in this setting. Also, we can add consistency as a further criterion to judge the quality of blends. As discussed, some relaxations of our algorithms (e.g. using bounded checks) may yield a better fit with human performance. We will also need to study more heuristics for the generalization relaxation stage, since they will affect the order in which optimal blends will be detected, and so the time needed to make the mathematically-oriented user satisfied by the produced blends.

Other algorithmic accounts are given, for instance, in [8], where the CB mechanism uses a parallel search engine based on genetic algorithms, or in [4], sketching the blending of logical theories within a distributed ontology setup. Further work on CB is contained in [6] where the authors present a rule-based system for counterfactual reasoning in natural language. These examples are mostly addressing problems from linguistics or philosophy, but our interest lies in particular in the blending of mathematical theories, as a means of understanding certain developments in the history of mathematics, as described by [1], and as part of general mathematical cognition, as suggested by [5].

Acknowledgements. T. R. Besold, K.-U. Kühnberger, A. Pease, and A. Smaill as members of the project COINVENT acknowledge the financial support of the Future and Emerging Technologies programme within the Seventh Framework Programme for Research of the European Commission, under FET-Open grant number: 611553.

References

1. Alexander, J.: Blending in Mathematics. Semiotica 2011(187), 1–48 (2011)
2. Fauconnier, G., Turner, M.: The Way We Think: Conceptual Blending and the Mind's Hidden Complexities. Basic Books, New York (2002)
3. Goguen, J.: Mathematical models of cognitive space and time. In: Andler, D., Ogawa, Y., Okada, M., Watanabe, S. (eds.) Reasoning and Cognition: Proc. of the Interdisciplinary Conference on Reasoning and Cognition, pp. 125–128. Keio University Press (2006)
4. Kutz, O., Bateman, J., Neuhaus, F., Mossakowski, T., Bhatt, M.: E pluribus unum: Formalisation, Use-Cases, and Computational Support for Conceptual Blending. In: Computational Creativity Research: Towards Creative Machines, Atlantis Thinking Machines, vol. 7. Springer (forthcoming in, 2015),
http://www.springer.com/computer/ai/book/978-94-6239-084-3
5. Lakoff, G., Núñez, R.: Where Mathematics Comes From: How the Embodied Mind Brings Mathematics into Being. Basic Books, New York (2000)
6. Lee, M., Barnden, J.: A Computational Approach to Conceptual Blending within Counterfactuals. Cognitive Science Research Papers CSRP-01-10, School of Computer Science, University of Birmingham (2001)

Decomposition of Some Jacobian Varieties of Dimension 3

Lubjana Beshaj and Tony Shaska

Dep. of Mathematics and Statistics, Oakland University, Rochester, MI, 48309

Abstract. We study degree 2 and 4 elliptic subcovers of hyperelliptic curves of genus 3 defined over $\mathbb{C}$. The family of genus 3 hyperelliptic curves which have a degree 2 cover to an elliptic curve E and degree 4 covers to elliptic curves E_1 and E_2 is a 2-dimensional subvariety of the hyperelliptic moduli $\mathcal{H}_3$. We determine this subvariety explicitly. For any given moduli point $\mathfrak{p} \in \mathcal{H}_3$ we determine explicitly if the corresponding genus 3 curve $\mathcal{X}$ belongs or not to such family. When it does, we can determine elliptic subcovers E, E_1, and E_2 in terms of the absolute invariants $t_1, \ldots, t_6$ as in [12]. This variety provides a new family of hyperelliptic curves of genus 3 for which the Jacobians completely split. The sublocus of such family when $E_1 \cong E_2$ is a 1-dimensional variety which we determine explicitly. We can also determine $\mathcal{X}$ and E starting form the j-invariant of E_1.

1   Introduction

There are some problems in classical mathematics which can be solved only through symbolic computational methods. The problem in which this work is focused lies within this category. Whether methods in artificial intelligence, machine learning etc can be improved to generalize such computational methods remains to be seen.

Let $\mathcal{M}_g$ denote the moduli space of genus $g \geq 2$ algebraic curves defined over an algebraically closed field k and $\mathcal{H}_g$ the hyperelliptic submoduli in $\mathcal{M}_g$. The sublocus of genus g hyperelliptic curves with an elliptic involution is a g-dimensional subvariety of $\mathcal{H}_g$. For $g = 2$ this space is denoted by $\mathcal{L}_2$ and studied in [11] and for $g = 3$ is denoted by $\mathcal{S}_2$ and is computed and discussed in detail in [4]. In both cases, a birational parametrization of these spaces is found via *dihedral invariants* which are introduced by the second author and generalized for any genus $g \geq 2$ in [6]. We denote the parameters for $\mathcal{L}_2$ by u, v and for $\mathcal{S}_2$ by $\mathfrak{s}_2, \mathfrak{s}_3, \mathfrak{s}_4$ as in respective papers. Hence, for the case $g = 3$ there is a birational map $\phi : \mathcal{S}_2 \longrightarrow \mathcal{H}_3$ such that $\phi : (\mathfrak{s}_2, \mathfrak{s}_3, \mathfrak{s}_4) = (t_1, \ldots, t_6)$, where $t_1, \ldots, t_6$ are the absolute invariants as defined in [12].

The dihedral invariants $\mathfrak{s}_2, \mathfrak{s}_3, \mathfrak{s}_4$ provide a birational parametrization of the locus $\mathcal{S}_2$. Hence, a generic curve in $\mathcal{S}_2$ is uniquely determined by the corresponding triple $(\mathfrak{s}_2, \mathfrak{s}_3, \mathfrak{s}_4)$. Let $\mathcal{X}$ be a curve in the locus $\mathcal{S}_2$. Then there is a degree 2 map $f_1 : \mathcal{X} \to E$ for some elliptic curve E. Thus, the Jacobian of $\mathcal{X}$ splits as Jac $(\mathcal{X}) \cong E \times A$, where A is a genus 2 Jacobian. Hence, there is a map

G.A. Aranda-Corral et al. (Eds.): AISC 2014, LNAI 8884, pp. 193–204, 2014.
© Springer International Publishing Switzerland 2014

$f_2 : \mathcal{X} \to C$ for some genus 2 curve C. The equations of $\mathcal{X}$, E, and C are given in Thm. 2. For any fixed curve $\mathcal{X} \in \mathcal{S}_2$, the subcovers E and C are uniquely determined in terms of the invariants $\mathfrak{s}_2, \mathfrak{s}_3, \mathfrak{s}_4$.

In section three we give the splitting of the Jacobians for all genus 3 algebraic curves, when this splitting is induced by automorphisms. The proof requires the Poincare duality and some basic group theory.

In this paper, we are mostly interested in the case when the Jacobian of the genus two curve C also splits. The Jacobian of C can split as an (n, n)-structure; see [8]. The loci of such genus 2 curves with $(3, 3)$-split or $(5, 5)$-split have been studied respectively in [7, 9]. For $n = 4$ the reader can check [5]. We focus on the case when the Jacobian of C is $(2, 2)$-split, which corresponds to the case when the Klein 4-group $V_4 \hookrightarrow \mathrm{Aut}\ (C)$. Hence, Jac $\mathcal{X}$ splits completely as a product of three elliptic curves. We say that Jac $\mathcal{X}$ is $(2, 4, 4)$-split.

Let the locus of genus 3 hyperelliptic curves whose Jacobian is $(2, 4, 4)$-split be denoted by $\mathcal{T}$. Then, there is a rational map $\psi : \mathcal{T} \to \mathcal{L}_2$ such that $\psi(\mathfrak{s}_2, \mathfrak{s}_3, \mathfrak{s}_4) = (u, v)$, which has degree 70 and can be explicitly computed, even though the rational expressions of u and v in terms of $\mathfrak{s}_2, \mathfrak{s}_3, \mathfrak{s}_4$ are quite large.

There are three components of $\mathcal{T}$ which we denote them by $\mathcal{T}_i$, $i = 1, 2, 3$. Two of these components are well known and the correspond to the cases when V_4 is embedded in the reduced automorphism group of $\mathcal{X}$. These cases correspond to the singular locus of $\mathcal{S}_2$ and are precisely the locus $\det\ (\mathrm{Jac}\ (\phi)) = 0$. This happens for all genus $g \geq 2$ as noted in [11]. The third component $\mathcal{T}_3$ is more interesting to us. It doesn't seem to have any group theoretic reason for this component to be there in the first place. We find the equation of this component it terms of the $\mathfrak{s}_2, \mathfrak{s}_3, \mathfrak{s}_4$ invariants. It is an equation $F_1(\mathfrak{s}_2, \mathfrak{s}_3, \mathfrak{s}_4) = 0$ as in Eq. (7). In this locus, the elliptic subfields of the genus two field $k(C)$ can be determined explicitly.

The main goal of this paper is to determine explicitly the family $\mathcal{T}_3$ of genus 3 curves and relations among its elliptic subcovers. We have the maps $\mathcal{T}_3 \xrightarrow{\psi} \mathcal{L}_2 \xrightarrow{\psi_0} k^2$, such that $\psi_0(\psi(\mathfrak{s}_2, \mathfrak{s}_3, \mathfrak{s}_4)) = (j_1, j_2)$, where $\mathfrak{s}_2, \mathfrak{s}_3, \mathfrak{s}_4$ satisfy Eq. (7) and u, v are given explicitly by Eq. (6) and Thm. (3) in [11]. The degree $\deg \psi_0 = 2$ and $\deg \psi = 70$.

Since $\mathcal{T}_3$ is a subvariety of $\mathcal{H}_3$ it would be desirable to express its equation in terms of a coordinate in $\mathcal{H}_3$. One can use the absolute invariants of the genus 3 hyperelliptic curves $t_1, \ldots, t_6$ as defined in [12] and the expressions of $\mathfrak{s}_2, \mathfrak{s}_3, \mathfrak{s}_4$ in terms of these invariants as computed in [4].

Further, we focus our attention to the sublocus $\mathcal{V}$ of $\mathcal{L}_2$ such that the genus 2 field $k(C)$ has isomorphic elliptic subfields. Such locus was discovered in [11] and it is somewhat surprising. It does not rise from a family of genus two curves with a fixed automorphism group as other families, see [11] for details. Using this sublocus of $\mathcal{M}_2$ we discover a rather unusual embedding $\mathcal{M}_1 \hookrightarrow \mathcal{M}_2$ as noted in [11]. Let $\mathfrak{T} \subset \mathcal{T}_3 \subset \mathcal{H}_3$ be the subvariety of $\mathcal{T}_3$ obtained by adding the condition

$j_1 = j_2$. Then, $\mathfrak{T}$ is a 1-dimensional variety defined by equations

$$
\begin{cases}
F_1(\mathfrak{s}_2, \mathfrak{s}_3, \mathfrak{s}_4) = 0 \\
F_2(\mathfrak{s}_2, \mathfrak{s}_3, \mathfrak{s}_4) = 0
\end{cases}
\tag{1}
$$

where F_2 is the discriminant of the quadratic polynomial roots of which are j-invariants j_1 and j_2; cf. Lemma 3. Hence, we have the maps $k \to \mathfrak{T} \hookrightarrow \mathcal{V} \hookrightarrow k$, such that $t \to (\mathfrak{s}_2, \mathfrak{s}_3, \mathfrak{s}_4) \to (u, v) \to j_1$.

Next we study whether the above maps are invertible. That would provide birational parameterizations for varieties $\mathcal{V}$ and $\mathfrak{T}$. The variety $\mathcal{V}$ is known to have a birational parametrization from Thm. (3) in [11]. The map can be inverted as $j \to (u, v) = \left(9 - \frac{j}{256}, 9\left(6 - \frac{j}{256}\right)\right)$; see [11] for details. The main computational task of this paper is to find a birational parametrization of $\mathfrak{T}$.

Given $(u, v) \in \mathcal{V}$ there is a unique (up to isomorphism) genus 2 curve C corresponding to this point in $\mathcal{V}$. From Lemma 3, every genus 2 curve can be written as in Eq. (11). Hence, there exists a triple $(\mathfrak{s}_2, \mathfrak{s}_3, \mathfrak{s}_4)$ corresponding to (u, v).

If $j \in \mathbb{Q}$, then the corresponding elliptic curve E_j is defined over $\mathbb{Q}$. From the above expressions we see that $u, v \in \mathbb{Q}$. Then, from [10] the corresponding genus two curve C has also minimal field of definition $\mathbb{Q}$. The same holds for $\mathfrak{s}_2, \mathfrak{s}_3, \mathfrak{s}_4$ and the genus 3 corresponding curve $\mathcal{X}$.

2 Decomposition of Jacobian Varieties of Dimension 3

Let $\mathcal{X}$ be a genus g algebraic curve with automorphism group $G := \text{Aut}(\mathcal{X})$. Let $H \leq G$ such that $H = H_1 \cup \cdots \cup H_t$ where the subgroups $H_i \leq H$ satisfy $H_i \cap H_j = \{1\}$ for all $i \neq j$. Then,

$$
\text{Jac}^{\,t-1}(\mathcal{X}) \times \text{Jac}^{\,|H|}(\mathcal{X}/H) \cong \text{Jac}^{\,|H_1|}(\mathcal{X}/H_1) \times \cdots \times \text{Jac}^{\,|H_t|}(\mathcal{X}/H_t)
$$

The group H satisfying these conditions is called a group with partition. Elementary abelian p-groups, the projective linear groups $PSL_2(q)$, Frobenius groups, dihedral groups are all groups with partition.

Let $H_1, \ldots, H_t \leq G$ be subgroups with $H_i \cdot H_j = H_j \cdot H_i$ for all $i, j \leq t$, and let g_{ij} denote the genus of the quotient curve $\mathcal{X}/(H_i \cdot H_j)$. Then, for $n_1, \ldots, n_t \in \mathbb{Z}$ the conditions $\sum n_i n_j g_{ij} = 0$, and $\sum_{j=1}^{t} n_j g_{ij} = 0$, imply the isogeny relation

$$
\prod_{n_i > 0} \text{Jac}^{\,n_i}(\mathcal{X}/H_i) \cong \prod_{n_j < 0} \text{Jac}^{\,|n_j|}(\mathcal{X}/H_j)
$$

In particular, if $g_{ij} = 0$ for $2 \leq i < j \leq t$ and if $g = g_{\mathcal{X}/H_2} + \cdots + g_{\mathcal{X}/H_t}$, then

$$
\text{Jac}(\mathcal{X}) \cong \text{Jac}(\mathcal{X}/H_2) \times \cdots \times \text{Jac}(\mathcal{X}/H_t)
$$

The reader can check [1] for the proof of the above statements.

2.1   Non-hyperelliptic Curves

We will use the above facts to decompose the Jacobians of genus 3 non-hyperelliptic curves. $\mathcal{X}$ denotes a genus 3 non-hyperelliptic curve unless otherwise stated and $\mathcal{X}_2$ denotes a genus 2 curve.

The Group C_2. Then the curve $\mathcal{X}$ has an elliptic involution $\mathfrak{s} \in$ Aut $(\mathcal{X})$. Hence, there is a Galois covering $\pi \colon \mathcal{X} \to \mathcal{X}/\langle\sigma\rangle =: E$. We can assume that this covering is maximal. The induced map $\pi^* \colon E \to$ Jac $(\mathcal{X})$ is injective. Then, the kernel projection Jac $(\mathcal{X}) \to E$ is a dimension 2 abelian variety. Hence, there is a genus 2 curve $\mathcal{X}_2$ such that Jac $(\mathcal{X}_2) \cong E \times$ Jac $(\mathcal{X}_2)$.

The Klein 4-group. Next, we focus on the automorphism groups G such that $V_4 \hookrightarrow G$. In this case, there are three elliptic involutions in V_4, namely $\sigma, \tau, \sigma\tau$. Obviously they form a partition. Hence, the Jacobian of $\mathcal{X}$ is the product Jac $^2(\mathcal{X}) \cong E_1^2 \times E_2^2 \times E_3^2$ of three elliptic curves. By applying the Poincare duality we get Jac $(\mathcal{X}) \cong E_1 \times E_2 \times E_3$.

The Dihedral Group D_8. In this case, we have 5 involutions in G in 3 conjugacy classes. No conjugacy class has three involutions. Hence, we can pick three involutions such that two of them are conjugate to each other in G and all three of them generate V_4. Hence, Jac $(\mathcal{X}) \cong E_1^2 \times E_2$, for some elliptic curves E_1, E_2.

The Symmetric Group S_4. The Jacobian of such curves splits into a product of elliptic curves since $V_4 \hookrightarrow S_4$. Below we give a direct proof of this.

We know that there are 9 involutions in S_4, six of which are transpositions. The other three are product of two 2-cycles and we denote them by $\sigma_1, \sigma_2, \sigma_3$. Let H_1, H_2, H_3 denote the subgroups generated by $\sigma_1, \sigma_2, \sigma_3$. They generate V_4 and are all isomorphic in G. Hence, Jac $(\mathcal{X}) \cong E^3$, for some elliptic curve E.

The Symmetric Group S_3. We know from above that the Jacobian is a direct product of three elliptic curves. Here we will show that two of those elliptic curves are isomorphic. Let H_1, H_2, H_3 be the subgroups generated by transpositions and H_4 the subgroup of order 3. Then

$$\text{Jac } ^3(\mathcal{X}) \cong E_1^2 \times E_2^2 \times E_3^2 \times \text{Jac } ^3(\mathcal{Y})$$

for three elliptic curves E_1, E_2, E_3 fixed by involutions and a curve $\mathcal{Y}$ fixed by the element of order 3. Simply by counting the dimensions we have $\mathcal{Y}$ to be another elliptic curve E_4. Since all the transpositions of S_3 are in the same conjugacy class then E_1, E_2, E_3 are isomorphic. Then by applying the Poincare duality we have that Jac $(X) \cong E^2 \times E'$.

Summarizing, we have the following:

Theorem 1. *Let $\mathcal{X}$ be a genus 3 curve and G its automorphism group. Then,*
a) If $\mathcal{X}$ is hyperelliptic, then the following hold:
*   i) If G is isomorphic to V_4 or $C_2 \times C_4$, then Jac (X) is isogenous to the product of an elliptic curve E and the Jacobian of a genus 2 curve $\mathcal{X}_2$, namely Jac $(\mathcal{X}) \cong E \times$ Jac $(\mathcal{X}_2)$.*

ii) If G is isomorphic to C_2^3 then Jac (X) is isogenous to the product of three elliptic curves, namely Jac $(\mathcal{X}) \cong E_1 \times E_2 \times E_3$.

iii) If G is isomorphic to $D_{12}, C_2 \times S_4$ or any of the groups of order 24 or 32, then Jac (X) is isogenous to the product of three elliptic curves such that two of them are isomorphic, namely Jac $(\mathcal{X}) \cong E_1^2 \times E_2$.

b) If $\mathcal{X}$ is non-hyperelliptic then the following hold:

i) If G is isomorphic to C_2, then Jac (X) is isogenous to the product of an elliptic curve and the Jacobian of some genus 2 curve $\mathcal{X}_2$, namely Jac $(\mathcal{X}) \cong E \times$ Jac $(\mathcal{X}_2)$.

ii) If G is isomorphic to V_4, then Jac (X) is isogenous to the product of three elliptic curves namely Jac $(\mathcal{X}) \cong E_1 \times E_2 \times E_3$.

iii) If G is isomorphic to S_3, D_8 or has order 16 or 48, then Jac (X) is isogenous to the product of three elliptic curves such that two of them are isomorphic to each other, namely Jac $(\mathcal{X}) \cong E_1^2 \times E_2$.

iv) If G is isomorphic to $S_4, L_3(2)$ or $C_2^3 \rtimes S_3$, then Jac (X) is isogenous to the product of three elliptic curves such that all three of them are isomorphic to each other, namely Jac $(\mathcal{X}) \cong E^3$.

Proof. The proof of the hyperelliptic case is similar and we skip the details.

Part b): When G is isomorphic to C_2, V_4, D_8, S_4, S_3 the result follows from the remarks above. The rest of the theorem is an immediate consequence of the list of groups as in the Table 1 of [6]. If $|G| = 16, 48$ then $D_8 \hookrightarrow G$. Then, from the remarks at the beginning of this section the results follows. If G is isomorphic to $L_3(2)$ or $C_4^2 \rtimes S_3$ then $S_4 \hookrightarrow G$. Hence the Jacobian splits as in the case of S_4. This completes the proof.   □

The above theorem gives the splitting of the Jacobian based on automorphisms. Next we will focus on the $(2, 4, 4)$ splitting for hyperelliptic curves. We will explicitly determine the elliptic components for a given genus 3 curve $\mathcal{X}$.

3   Hyperelliptic Curves with Extra Involutions

Let K be a genus 3 hyperelliptic field over the ground field k. Then K has exactly one genus 0 subfield of degree 2, call it $k(X)$. It is the fixed field of the **hyperelliptic involution** ω_0 in Aut (K). Thus, ω_0 is central in Aut (K), where Aut (K) denotes the group Aut (K/k). It induces a subgroup of Aut $(k(X))$ which is naturally isomorphic to $\overline{\text{Aut}}(K) := \text{Aut}(K)/\langle \omega_0 \rangle$. The latter is called the **reduced automorphism group** of K.

An **elliptic involution** of $G = \text{Aut}(K)$ is an involution which fixes an elliptic subfield. An involution of $\bar{G} = \overline{\text{Aut}}(K)$ is called **elliptic** if it is the image of an elliptic involution of G. If ω_1 is an elliptic involution in G then $\omega_2 := \omega_0 \omega_1$ is another involution (not necessarily elliptic). So the non-hyperelliptic involutions come naturally in (unordered) pairs ω_1, ω_2. These pairs correspond bijectively to the Klein 4-groups in G.

Definition 1. *We will consider pairs (K, ε) with K a genus 3 hyperelliptic field and ε an elliptic involution in $\bar{G}$. Two such pairs (K, ε) and (K', ε') are called isomorphic if there is a k-isomorphism $\alpha : K \to K'$ with $\varepsilon' = \alpha \varepsilon \alpha^{-1}$.*

Let ε be an elliptic involution in $\bar{G}$. We can choose the generator X of $\mathrm{Fix}(\omega_0)$ such that $\varepsilon(X) = -X$. Then $K = k(X, Y)$ where X, Y satisfy equation $Y^2 = (X^2 - \alpha_1^2)(X^2 - \alpha_2^2)(X^2 - \alpha_3^2)(X^2 - \alpha_4^2)$, for some $\alpha_i \in k$, $i = 1, \ldots, 4$. Denote by s_1, s_2, s_3, s_4 the symmetric polynomials of $\alpha_1^2, \alpha_2^2, \alpha_3^2, \alpha_4^2$; see [15] for details. Then, we have $Y^2 = X^8 + s_1 X^6 + s_2 X^4 + s_3 X^2 + s_4$, with $s_1, s_2, s_3, s_4 \in k$, $s_4 \neq 0$. Furthermore, $E = k(X^2, Y)$ and $C = k(X^2, YX)$ are the two subfields corresponding to ε of genus 1 and 2 respectively.

Preserving the condition $\varepsilon(X) = -X$ we can further modify X such that $s_4 = 1$. Then, we have the following:

Theorem 2. *Let K be a genus 3 hyperelliptic field and F an elliptic subfield of degree 2.*

i) Then, $K = k(X, Y)$ such that

$$Y^2 = X^8 + aX^6 + bX^4 + cX^2 + 1 \tag{2}$$

for $a, b, c \in k$ such that the discriminant of the right hand side $\Delta(a, b) \neq 0$.

ii) $F = k(U, V)$ where $U = X^2$, and $V = Y$ and

$$V^2 = U^4 + aU^3 + bU^2 + cU + 1 \tag{3}$$

iii) There is a genus 2 subfield $L = k(x, y)$ where $x = X^2$, $y = YX$ and

$$y^2 = x(x^4 + ax^3 + bx^2 + cx + 1) \tag{4}$$

Proof. The proof follows from the above remarks. To show that the genus 2 subfield is generated by X^2, YX it is enough to show that they are fixed by ω_2. In cases ii) and iii) we are again assuming that the discriminant of the right hand side is not zero.   □

These conditions determine X up to coordinate change by the group $\langle \tau_1, \tau_2 \rangle$ where $\tau_1 : X \to \zeta_8 X$, $\tau_2 : X \to \frac{1}{X}$, and ζ_8 is a primitive 8-th root of unity in k. Hence, $\tau_1 : (a, b, c) \to (\zeta_8^6 a, \zeta_8^4 b, \zeta^2 c)$, and $\tau_2 : (a, b, c) \to (c, b, a)$.

Then, $|\tau_1| = 4$ and $|\tau_2| = 2$. The group generated by τ_1 and τ_2 is the dihedral group of order 8. Invariants of this action are

$$s_2 = ac, \quad s_3 = (a^2 + c^2)b, \quad s_4 = a^4 + c^4. \tag{5}$$

Since the above transformations are automorphisms of the projective line $\mathbb{P}^1(k)$ then the $SL_2(k)$ invariants must be expressed in terms of s_2, s_3, and s_4.

If $s_4 + 2s_2^2 = 0$ then this implies that the curve has automorphism group $\mathbb{Z}_2 \times \mathbb{Z}_4$, see [4] for details. From now on we assume that $s_4 + 2s_2^2 \neq 0$.

The discriminant of the octavic polynomial on the right hand side of Eq. (2) is expressed in terms of s_2, s_3, s_4; see [15] for details. From now forward we will assume that $\Delta(s_2, s_3, s_4) \neq 0$ since in this case the corresponding triple (s_2, s_3, s_4) does not correspond to a genus 3 curve. The map $(a, b, c) \mapsto (s_2, s_3, s_4)$ is a branched Galois covering with group D_4 of the set $\{(s_2, s_3, s_4) \in k^3 : \Delta(s_2, s_3, s_4) \neq 0\}$ by the corresponding open subset of (a, b, c)-space. In any case, it is true that if a, b, c and a', b', c' have the same s_2, s_3, s_4-invariants then they are conjugate under $\langle \tau_1, \tau_2 \rangle$.

Lemma 1. *For $(a,b,c) \in k^3$ with $\Delta \neq 0$, equation (2) defines a genus 3 hyperelliptic field $K_{a,b,c} = k(X,Y)$. Its reduced automorphism group contains the non-hyperelliptic involution $\varepsilon_{a,b,c} : X \mapsto -X$. Two such pairs $(K_{a,b,c}, \varepsilon_{a,b,c})$ and $(K_{a',b',c'}, \varepsilon_{a',b',c''})$ are isomorphic if and only if $\mathfrak{s}_4 = \mathfrak{s}'_4$, $\mathfrak{s}_3 = \mathfrak{s}'_3$, and $\mathfrak{s}_2 = \mathfrak{s}'_2$, where $\mathfrak{s}_4, \mathfrak{s}_3, \mathfrak{s}_2$ and $\mathfrak{s}'_4, \mathfrak{s}'_3, \mathfrak{s}'_2$ are associated with a,b,c and a',b',c', respectively, by (5)).*

Proof. An isomorphism α between these two pairs yields $K = k(X,Y) = k(X',Y')$ with $k(X) = k(X')$ such that X,Y satisfy (2) and X',Y' satisfy the corresponding equation with a,b,c replaced by a',b',c'. Further, $\varepsilon_{a,b,c}(X') = -X'$. Thus X' is conjugate to X under $\langle \tau_1, \tau_2 \rangle$ by the above remarks. This proves the condition is necessary. It is clearly sufficient. $\qquad \square$

Relations among $\mathfrak{s}_2, \mathfrak{s}_3, \mathfrak{s}_4$ for each G when $V_4 \hookrightarrow G$ are determined in [4].

4   Subcovers of Genus 2

Next we study in detail the complement C of E in Jac $(\mathcal{X})$. From the above theorem, C has equation as in Eq. (4). Its absolute invariants i_1, i_2, i_3, as defined in [11], can be expressed in terms of the dihedral invariants $\mathfrak{s}_4, \mathfrak{s}_3, \mathfrak{s}_2$ as follows:

$$i_1 = 144 \frac{M}{D^2} f_1(\mathfrak{s}_2, \mathfrak{s}_3, \mathfrak{s}_4), \quad i_2 = 432 \frac{M^2}{D^3} f_2(\mathfrak{s}_2, \mathfrak{s}_3, \mathfrak{s}_4), \quad i_3 = \frac{243}{16} \frac{M^3}{D^5} f_3(\mathfrak{s}_2, \mathfrak{s}_3, \mathfrak{s}_4) \quad (6)$$

where $M = \mathfrak{s}_4 + 2\mathfrak{s}_2^2$ and $D = 16\,\mathfrak{s}_2{}^3 - 40\,\mathfrak{s}_2{}^2 + 8\,\mathfrak{s}_2\mathfrak{s}_4 - 3\,\mathfrak{s}_3{}^2 - 20\,\mathfrak{s}_4$ and f_1, f_2, f_3 are given in [15]. For the rest of the paper we assume that $D = J_2 \neq 0$.

We consider the case when Jac (C) is $(2,2)$ decomposable. The locus $\mathcal{L}_2$ of such genus two curves is computed in [11] in terms of the invariants i_1, i_2, i_3. Substituting the expressions in Eq. (6) in the equation of $\mathcal{L}_2$ from [11] we have the following:

$$\left(2\,\mathfrak{s}_2{}^2 - \mathfrak{s}_4\right) \cdot \left(2\,\mathfrak{s}_2{}^2 + \mathfrak{s}_4\right) \cdot F_1(\mathfrak{s}_2, \mathfrak{s}_3, \mathfrak{s}_4) = 0 \quad (7)$$

where $F_1(\mathfrak{s}_2, \mathfrak{s}_3, \mathfrak{s}_4)$ is an irreducible polynomial of degree 13, 8, 6 in $\mathfrak{s}_2, \mathfrak{s}_3, \mathfrak{s}_4$ respectively; see [15].

Let the locus of genus 3 hyperelliptic curves whose Jacobian is $(2,4,4)$-split be denoted by $\mathcal{T}$. There are three components of $\mathcal{T}$ which we denote them by $\mathcal{T}_i$, $i = 1,2,3$ as seen by Eq. (7).

Two of these components are well known and the correspond to the cases when V_4 is embedded in the reduced automorphism group of $\mathcal{X}$. These cases correspond to the singular locus of $\mathcal{S}_2$ and are precisely the locus det (Jac (ϕ)) = 0, see [11]. This happens for all genus $g \geq 2$.

Lemma 2. *Let $\mathcal{X}$ be a genus 3 curve with $(2,2,4)$-split Jacobian. Then, one of the following occurs*
  i) $\mathbb{Z}_2^3 \hookrightarrow$ Aut $(\mathcal{X})$
  ii) $\mathbb{Z}_2 \times \mathbb{Z}_4 \hookrightarrow$ Aut $(\mathcal{X})$
  iii) $\mathcal{X}$ is in the locus $\mathcal{T}_3$

Proof. The proof is an immediate consequence of Theorem 2 and Eq. (7).   □

The third component $\mathcal{T}_3$ is more interesting to us. It is the moduli space of pairs of degree 4 non-Galois covers $\psi_i : \mathcal{X}_3 \to E_i$, $i = 1, 2$.

One of the main goals of this paper is to determine explicitly the family $\mathcal{T}_3$ of genus 3 curves and relations among its elliptic subcovers. We have the maps

$$\mathcal{T}_3 \xrightarrow{\psi} \mathcal{L}_2 \xrightarrow{\psi_0} k^2 \tag{8}$$
$$(\mathfrak{s}_2, \mathfrak{s}_3, \mathfrak{s}_4) \to (u, v) \to (j_1, j_2)$$

where $\mathfrak{s}_2, \mathfrak{s}_3, \mathfrak{s}_4$ satisfy $F_1(\mathfrak{s}_2, \mathfrak{s}_3, \mathfrak{s}_4) = 0$ and u, v are given explicitly by Eq. (6) and Thm. (3) in [11].

The expressions of u and v are computed explicitly in terms of $\mathfrak{s}_2, \mathfrak{s}_3, \mathfrak{s}_4$ by substituting the expressions of Eq. (6) expressions for u and v as rational functions of $i_1, i_2 i_3$ as computed in [11]. As rational functions u and v have degrees 35 and 70 respectively (in terms of $\mathfrak{s}_2, \mathfrak{s}_3, \mathfrak{s}_4$). The j-invariants j_1 and j_2 will be determined in the next section.

Since $\mathcal{T}_3$ is a subvariety of $\mathcal{H}_3$ it would be desirable to express its equation in terms of a coordinate in $\mathcal{H}_3$. One can use the absolute invariants of the genus 3 hyperelliptic curves $t_1, \ldots, t_6$ as defined in [12] and the expressions of $\mathfrak{s}_4, \mathfrak{s}_3, \mathfrak{s}_2$ in terms of these invariants as computed in [4].

Remark 1. $\mathcal{T}_3$ *is a 2-dimensional subvariety of* $\mathcal{H}_3$ *determined by the equations*

$$\begin{cases} F_1(\mathfrak{s}_2, \mathfrak{s}_3, \mathfrak{s}_4) = 0 \\ t_i - T_i(\mathfrak{s}_2, \mathfrak{s}_3, \mathfrak{s}_4), \quad i = 1, \ldots, 6 \end{cases} \tag{9}$$

where T_i is the function t_i evaluated for the triple $(\mathfrak{s}_2, \mathfrak{s}_3, \mathfrak{s}_4)$.

The equations of $\mathcal{T}_3$ can be explicitly determined in terms of $t_1, \ldots, t_6$ by eliminating $\mathfrak{s}_2, \mathfrak{s}_3, \mathfrak{s}_2$ from the above equations. Normally, when we talk about $\mathcal{T}_3$ we will think of it given in terms of $t_1, \ldots, t_6$.

Example 1. *Consider the genus 3 curves $\mathcal{X}$ with* $\mathrm{Aut}\,(\mathcal{X}) \cong \mathbb{Z}_2^3$. *Then,* $\mathfrak{s}_4 = 2 s_2^2$ *and*

$$u = \frac{1}{P}\left(-9\,s_3{}^2 + 120\,s_2 s_3 - 400\,s_2{}^2 + 16\,s_2{}^3\right)$$

$$v = -\frac{2}{P^2}\left(432\,s_2 s_3{}^3 - 27\,s_3{}^4 - 1440\,s_2{}^2 s_3{}^2 - 6400\,s_2{}^3 s_3 + 32000\,s_2{}^4\right.$$
$$\left. + 288\,s_2{}^3 s_3{}^2 - 5376\,s_2{}^4 s_3 + 23040\,s_2{}^5 + 256\,s_2{}^6\right)$$

where $P = -s_3^2 - 8\,s_2 s_3 - 16\,s_2^2 + 16\,s_2^3$.

For the rest of this section we will see if we can invert the map ψ.

Proposition 1. *Let $(u, v) \in k^2$ such that*

$$(u^2 - 4v + 18u - 27)(v^2 - 4u^3)(4v - u^2 + 110u - 1125) \neq 0.$$

Then, the curve of genus 2 defined over k given by

$$y^2 = a_0 x^6 + a_1 x^5 + a_2 x^4 + a_3 x^3 + t a_2 x^2 + t^2 a_1 x + t^3 a_0, \qquad (10)$$

corresponds to the moduli point $(u,v) \in \mathcal{L}_2 \hookrightarrow \mathcal{M}_2$, where one of the following holds:

i) If $u \neq 0$, then $t = v^2 - 4u^3$, $a_0 = v^2 + u^2 v - 2u^3$, $a_1 = 2(u^2 + 3v)(v^2 - 4u^3)$, $a_2 = (15v^2 - u^2 v - 30u^3)(v^2 - 4u^3)$, and $a_3 = 4(5v - u^2)(v^2 - 4u^3)^2$.

ii) If $u = 0$, then $t = 1$, $a_0 = 1 + 2v$, $a_1 = 2(3 - 4v)$, $a_2 = 15 + 14v$, $a_3 = 4(5 - 4v)$.

Hence, corresponding to the pair (u,v) there is a unique genus 2 curve $C_{u,v}$. The following Lemma addresses the rest of our question.

Lemma 3. *i) Any genus two curve C defined over an algebraically closed field k can be written as*

$$y^2 = x(x^4 + ax^3 + bx^2 + cx + 1) \qquad (11)$$

for some $a, b, c \in k$ such that $\Delta(a,b,c) \neq 0$.

ii) Let C be a genus 2 curve with equation as in Eq. (11). Then, there exists a genus 3 curve $\mathcal{X}$ with equation $Y^2 = X^8 + aX^6 + bX^4 + CX^2 + 1$ and a degree 2 map $f : \mathcal{X} \to C$ such that $x = X^2$ and $y = YX$.

Proof. i) Let C be a genus 2 curve defined over k. Then, the equation of C is given by $y^2 = \Pi_{i=1}^{6}(x - \alpha_i)$, where α_i are all distinct for all $i = 1 \ldots 6$. Since k is algebraically closed, then we can pick a change of transformation in $\mathbb{P}^1(k)$ such that $\alpha_1 \to 0$ and $\alpha_2 \to \infty$. We can also pick a coordinate such that $\alpha_3 \cdots \alpha_6 = 1$. Then, the curve C has equation as claimed. The condition that $\Delta(a,b,c) \neq 0$ simply assures that not two roots of the sextic coalesce.

ii) This genus 3 curve is a covering of C from Thm. 2.   □

Hence, the curve $C_{u,v}$ can be written as in Eq. (11). This would mean that we can explicitly compute $\mathfrak{s}_4, \mathfrak{s}_3, \mathfrak{s}_2$ in terms of u and v. Finding a general formula for $(\mathfrak{s}_2, \mathfrak{s}_3, \mathfrak{s}_4)$ in terms of (u,v) is computationally difficult. Under some additional restrictions this ca be done, as we will see in the next section.

5   Elliptic Subfields

In this section we will determine the elliptic subcovers of the genus 3 curves $\mathcal{X} \in \mathcal{T}_3$. We will describe how this can be explicitly done, but will skip displaying the computations here. A point $\mathfrak{p} = (t_1, \ldots, t_6) \in \mathcal{T}_3$ satisfies equations Eq. (9). Our goal is to determine the j-invariants of E, E_1, E_2 in terms of $t_1, \ldots t_6$. The j-invariant of E is

$$j = 256 \frac{\left(-\mathfrak{s}_3{}^2 - 12\,\mathfrak{s}_4 - 24\,\mathfrak{s}_2{}^2 + 3\,\mathfrak{s}_2\mathfrak{s}_4 + 6\,\mathfrak{s}_2{}^3\right)^3}{\left(\mathfrak{s}_4 + 2\,\mathfrak{s}_2{}^2\right)} f(\mathfrak{s}_2, \mathfrak{s}_3, \mathfrak{s}_4), \qquad (12)$$

where $f(\mathfrak{s}_2, \mathfrak{s}_3, \mathfrak{s}_4)$ can be found in [15].

We denote the degree 2 elliptic subcovers of C by E_1 and E_2 and their j-invariants by j_1 and j_2. These j invariants are the roots of the quadratic

$$j^2 + 256 \frac{2u^3 - 54u^2 + 9uv - v^2 + 27v}{u^2 + 18u - 4v - 27} j + 65536 \frac{u^2 + 9u - 3v}{(u^2 + 18u - 4v - 27)^2}, \quad (13)$$

see Eq. (4) in [11].

Since these j-invariants are determined explicitly in terms of u and v, then via the map $\psi : \mathcal{T}_3 \to \mathcal{L}_2$ we express such coefficients in terms of $\mathfrak{s}_2, \mathfrak{s}_3, \mathfrak{s}_4$. Moreover, the maps

$$\mathcal{T}_3 \to k^3 \setminus \{\Delta(\mathfrak{s}_2, \mathfrak{s}_3, \mathfrak{s}_4) = 0\} \to k^2 \setminus \{\Delta_{u,v} = 0\} \to k^2$$

$$(t_1, \ldots, t_6) \to (\mathfrak{s}_4, \mathfrak{s}_3, \mathfrak{s}_2) \xrightarrow{\psi} (u, v) \to (j_1, j_2) \quad (14)$$

are all explicitly determined.

Example 2. *Let be given a 6-tuple*

$$(t_1, \ldots, t_6) = \left(\frac{8767521}{6224272}, \frac{152464}{5329}, \frac{8116}{3431}, -\frac{3343532}{695617}, -\frac{91532}{148117}, -\frac{50448727768}{28398241} \right)$$

which satisfies the Eq. (17) in [12]. Then, this tuple corresponds to a genus 3 hyperelliptic curve, more precisely the curve $\mathcal{X}$ with equation

$$Y^2 = X^8 + X^6 + X^4 + X^2 + 1$$

Then, the corresponding invariants are $\mathfrak{s}_2 = 1$, $\mathfrak{s}_3 = 2$, $\mathfrak{s}_4 = 2$. The genus 2 subcover has invariants $i_1 = -\frac{48}{5}$, $i_2 = \frac{432}{5}$, $i_3 = \frac{1}{400}$ and the corresponding dihedral invariants are $u = 9$ and $v = -\frac{754}{5}$. The j-invariants of the three elliptic subcover are $j = 2048$ $j_1 = \frac{32768}{5} + \frac{2}{5} \sqrt{268435081}$ and $j_2 = \frac{32768}{5} - \frac{2}{5} \sqrt{268435081}$.

Next we will study the subvariety of $\mathcal{T}_3$, such that E_1 is isomorphic to E_2.

5.1   Isomorphic Elliptic Subfields

The two elliptic curves E_1 and E_2 are isomorphic when their j-invariants are equal, which happens when the discriminant of the quadratic in Eq. (13) is zero. From Remark (1) in [11] this occurs if and only if

$$(v^2 - 4u^3)(v - 9u + 27) = 0$$

The first condition is equivalent to $D_8 \hookrightarrow Aut(C)$. The later condition gives $u = 9 - \frac{\lambda}{256}$ and $v = 9 \left(6 - \frac{\lambda}{256} \right)$, where $\lambda := j_1 = j_2$. Both of these loci can be explicitly computed given enough computing power.

Substituting u and v in terms of $\mathfrak{s}_2, \mathfrak{s}_3, \mathfrak{s}_4$ in the equation $v - 9u + 27 = 0$, we get an equation of degree 68, 42, and 29 in $\mathfrak{s}_2, \mathfrak{s}_3, \mathfrak{s}_4$ respectively. We denote it by

$$F_2(\mathfrak{s}_2, \mathfrak{s}_3, \mathfrak{s}_4) = 0 \quad (15)$$

and do not display it here because of its size. This equation and the Eq. (7) define the locus $\mathfrak{T}$ in terms of $\mathfrak{s}_4, \mathfrak{s}_3, \mathfrak{s}_2$.

Lemma 4. *The algebraic variety $\mathfrak{T}$ is a 1-dimensional subvariety, it has 5 genus 0 components as in Eq. (16). Every point $(\mathfrak{s}_2, \mathfrak{s}_3, \mathfrak{s}_4) \in \mathfrak{T}$ correspond to a genus 3 hyperelliptic curve with $(2, 4, 4)$-split Jacobian such that the degree 4 elliptic subcovers are isomorphic to each other.*

Proof. From the equations above we can eliminate $\mathfrak{s}_3$ via resultants and get the following. In this case we get

$$(2\mathfrak{s}_2^2 - \mathfrak{s}_4)^{16}(\mathfrak{s}_4 + 2\mathfrak{s}_2^2)^{172}\, g_1^{12}\, g_2^{12}\, g_3^{10}\, g_4^8\, g_5 = 0 \tag{16}$$

where g_5 can be found in [15] and $g_1, \dots, g_4$ are

$$g_1 = s_4 + 2\, s_2{}^2 - 100\, s_2 + 625$$

$$g_2 = -27\, s_4 + s_2{}^3 + 6\, s_2{}^2 + 768\, s_2 - 4096$$

$$g_3 = -16777216 + 5242880\, s_2 - 450560\, s_2{}^2 + 7680\, s_2{}^3 - 340\, s_2{}^4 + 8\, s_2{}^5 - 102400\, s_4$$
$$+ 16640\, s_2 s_4 - 220\, s_2{}^2 s_4 + 4\, s_4 s_2{}^3 - 125\, s_4{}^2$$

$$g_4 = 3515625 - 937500\, s_2 + 62500\, s_2{}^2 + 64\, s_2{}^4 + 15000\, s_4 - 2000\, s_2 s_4$$

Since $(2\mathfrak{s}_4 - \mathfrak{s}_2^2)(2\mathfrak{s}_4 + \mathfrak{s}_2^2) \neq 0$, as noted before. All other components are genus zero curves.   □

The equation of $\mathfrak{T}$ can be expressed in the absolute invariants $t_1, \dots, t_6$ by eliminating $\mathfrak{s}_4, \mathfrak{s}_3, \mathfrak{s}_2$ from expressions in Eq. (6). Such expressions are large and we do not display them here.

Let be given a parameterization of $\mathfrak{T}$. Then we have the following maps

$$k \to \mathfrak{T} \to L_2 \to k$$
$$t \to (\mathfrak{s}_4(t), \mathfrak{s}_3(t), \mathfrak{s}_2(t)) \to (u(t), v(t)) \to j(t) \tag{17}$$

This map gives us the possibility to construct a family of curves defined over $\mathbb{Q}$ such that all their subcovers, namely C, E, E_1, and E_2 are also defined over $\mathbb{Q}$. For example, for $t \in \mathbb{Q}$ we have the corresponding $\mathfrak{s}_4, \mathfrak{s}_3, \mathfrak{s}_3 \in \mathbb{Q}$. Hence, there is a genus 3 curve $\mathcal{X}$ defined over $\mathbb{Q}$. The invariants u, v are rational functions of $\mathfrak{s}_4, \mathfrak{s}_3, \mathfrak{s}_2$ and therefore of t. Thus, $u, v \in \mathbb{Q}$. Form Prop. 1 there is a genus 2 curve C such that C is defined over $\mathbb{Q}$. Moreover, the j-invariants for all elliptic subcovers are rational functions in $\mathfrak{s}_4, \mathfrak{s}_3, \mathfrak{s}_2$ and therefore in t. Hence, E, E_1, E_2 are also defined over $\mathbb{Q}$.

Theorem 3. *Let $\mathcal{X}$ be a curve in $\mathfrak{T}$ and $\mathfrak{s}_2, \mathfrak{s}_3, \mathfrak{s}_4$ its corresponding dihedral invariants. Then*

$$\mathrm{Jac}\,(\mathcal{X}) \cong E \times E' \times E'$$

where E and E' are elliptic curves with j-invariants $j(E)$ as in Eq. (12) and $j(E')$ as

$$j' = -128\,\frac{2\,u^3 - 54\,u^2 + 9\,uv - v^2 + 27\,v}{u^2 + 18\,u - 4\,v - 27},$$

where u and v are given as rational functions of i_1, i_2, i_3 as in [11]. Moreover, there is only a finite number of genus 3 curves $\mathcal{X}$ such that $E \cong E'$.

Proof. The equation of $j(E)$ was computed in Eq. (12). Since the other two elliptic subcovers have the same j-invariants then this invariant is given by the double root of the quadratic in Eq. (13). Thus,

$$j' = -128 \frac{2\,u^3 - 54\,u^2 + 9\,uv - v^2 + 27\,v}{u^2 + 18\,u - 4\,v - 27}$$

Substituting the values for u and v we get the expression as claimed.

We have $E \cong E'$ if and only if $j = j'$. This gives a third equation $G(\mathfrak{s}_2, \mathfrak{s}_3, \mathfrak{s}_4) = 0$ as claimed in the theorem. By Bezut's theorem, the number of solutions of the system of equations $F_i(\mathfrak{s}_2, \mathfrak{s}_3, \mathfrak{s}_4) = 0$, for $i = 1, 2$ and $G(\mathfrak{s}_2, \mathfrak{s}_3, \mathfrak{s}_4) = 0$ is finite.   □

The family above could be significant in number theory in constructing genus 3 curves with many rational points. Similar techniques have been used for genus 2 in [10] and by various other authors [2, 3, 14].

References

[1] Accola, R.D.M.: Two theorems on Riemann surfaces with noncyclic automorphism groups. Proc. Amer. Math. Soc. 25, 598–602 (1970)

[2] Beshaj, L., Shaska, T.: Heights on algebraic curves, arXiv preprint arXiv:1406.5659 (2014)

[3] Beshaj, L., Shaska, T., Shor, C.: On Jacobians of curves with superelliptic components, Contemporary Mathematics, arXiv:1310.7241[math.AG] (to appear, 2015)

[4] Beshaj, L., Thompson, F.: Equations for superelliptic curves over their minimaleld of definition. Albanian J. Math. 8(1), 3–8 (2014)

[5] Bruin, N., Doerksen, K.: The arithmetic of genus two curves with (4; 4)-split Jacobians. Canad. J. Math. 63(5), 992–1024 (2011)

[6] Gutierrez, J., Shaska, T.: Hyperelliptic curves with extra involutions. LMS J. Comput. Math. 8, 102–115 (2005)

[7] Magaard, K., Shaska, T., Völklein, H.: Genus 2 curves that admit a degree 5 map to an elliptic curve. Forum Math 21(3), 547–566 (2009)

[8] Shaska, T.: Curves of genus 2 with (n; n)-decomposable Jacobians. J. Symbolic Comput. 31(5), 603–617 (2001)

[9] Shaska, T.: Genus 2 fields with degree 3 elliptic subfields. Forum Math 16(2), 263–280 (2004)

[10] Shaska, T.: Genus 2 Curves with (3,3)-Split Jacobian and Large Automorphism Group. In: Fieker, C., Kohel, D.R. (eds.) ANTS 2002. LNCS, vol. 2369, pp. 205–218. Springer, Heidelberg (2002)

[11] Shaska, T., Beshaj, L.: The arithmetic of genus two curves, Information security, coding theory and related combinatorics. NATO Sci. Peace Secur. Ser. D Inf. Commun. Secur., vol. 29, pp. 59–98. IOS, Amsterdam (2011)

[12] Shaska, T.: Some Remarks on the Hyperelliptic Moduli of Genus 3, Comm. Algebra 42(9), 4110–4130 (2014)

[13] Shaska, T., Thompson, F.: Bielliptic curves of genus 3 in the hyperelliptic moduli. Appl. Algebra Engrg. Comm. Comput. 24(5), 387–412 (2013)

[14] Shaska, T.: Families of genus 2 curves with many elliptic subcovers, arXiv:1209.0434v1

[15] Shaska, T.: Genus 3 hyperelliptic curves with (2, 4, 4) decomposable Jacobians, arXiv:1306.5284 [math.AG]

Author Index